Nils Posse

The Swedish System of Educational Gymnastics

Nils Posse

The Swedish System of Educational Gymnastics

ISBN/EAN: 9783337034337

Printed in Europe, USA, Canada, Australia, Japan

Cover: Foto ©berggeist007 / pixelio.de

More available books at **www.hansebooks.com**

THE SWEDISH SYSTEM

OF

EDUCATIONAL GYMNASTICS

BY

Baron NILS POSSE

GRADUATE OF THE ROYAL GYMNASTIC CENTRAL INSTITUTE STOCKHOLM SWEDEN
LATE 2D LIEUTENANT 1ST REGT. ROYAL SWEDISH FIELD ARTILLERY
FORMERLY INSTRUCTOR IN THE STOCKHOLM GYMNASTIC AND FENCING CLUB ETC

TWO HUNDRED AND FORTY-ONE ILLUSTRATIONS

BOSTON MDCCCXC
LEE AND SHEPARD PUBLISHERS
10 MILK STREET NEXT "OLD SOUTH MEETING HOUSE"
NEW YORK CHARLES T. DILLINGHAM
718 AND 720 BROADWAY

DEDICATED TO

Col. Gustaf Nyblaeus K.C.S.
LATE DIRECTOR ROYAL GYMNASTIC CENTRAL INSTITUTE STOCKHOLM SWEDEN

AND TO

Prof. L. M. Törngren K.S. K.G.V.
DIRECTOR ROYAL GYMNASTIC CENTRAL INSTITUTE STOCKHOLM SWEDEN

PREFACE.

THE total absence in the English language of a comprehensive handbook of Swedish gymnastics has induced the author to prepare this little treatise, the chief aim of which was originally to meet the urgent demand from the author's many pupils in the "Boston School of Gymnastics" for a text-book on the subject. At the same time, the author has tried to arrange the book so that professional teachers, already familiar with other forms of gymnastics, and possibly also an interested public, might find it useful in case they should desire to become acquainted with this much-talked-of system, or to apply it practically.

The author prefers to call this the Swedish system, although it was originated by P. H. Ling; for many improvements have been made since Ling's day, and what he devised has been changed from a personal into a national matter, just as has Jahn's movement in Germany.

Although, for reference, use has been made of the scant literature on this subject, yet the author wishes to state that this book is not a translation.

The chapters on the physiological effects of exercise and on the muscular activity in the bodily movements are not strictly within the scope of the book, yet the author has decided to include them in the appendix, that the teacher of gymnastics may always have them at hand for reference.

Part of the illustrations are original, but the majority are copies of drawings by the late Prof. Hj. Ling, which have been published in his "*Allmän Rörelselära*" (Stockholm, 1880) and in Prof. Törngren's "*Atlas till gymnastik-reglemente*" (Stockholm, 1879), the author having acquired the copyright of those used.

203 HUNTINGTON AVENUE, BOSTON.
    January, 1890.

CONTENTS

Part I
GENERAL THEORIES
	PAGE
DEFINITIONS	1
CHOICE OF EXERCISES	2
PROGRESSION	4
APPARATUS	8
WORDS OF COMMAND	19
GYMNASTIC DRESS	25

Part II
MOVEMENTS OF THE SWEDISH SYSTEM OF GYMNASTICS
DEFINITIONS AND GENERAL THEORIES	29
INTRODUCTORY EXERCISES	33
LEG MOVEMENTS	46
ARCH FLEXIONS	71
HEAVING MOVEMENTS	87
SHOULDER-BLADE MOVEMENTS	120
ABDOMINAL EXERCISES	137
LATERAL TRUNK MOVEMENTS	151
JUMPING AND VAULTING	168
RESPIRATORY EXERCISES	199
RULES FOR MAKING TABLES OF EXERCISES	208

Part III
APPENDIX
MEDICO-GYMNASTIC EXERCISES USED IN EDUCATIONAL GYMNASTICS	215
RULES FOR MEASURING PUPILS	222
PREPARATORY SWIMMING EXERCISES	227
PHYSIOLOGICAL EFFECTS OF EXERCISE	230
MEMORANDA OF THE MUSCULAR ACTIVITY IN THE PRINCIPAL MOVEMENTS OF THE HUMAN BODY	241
TABLES OF EXERCISES	256

EDUCATIONAL GYMNASTICS

PART I

GENERAL THEORIES

DEFINITIONS.

THE WORD **Gymnastics** is generally understood as meaning systematic exercise of the muscles for the restoration of health, and for the development and preservation of the physical powers. According to the purposes for which they are used, gymnastics might now be divided into two main branches: viz., *medical gymnastics*, or exercise for the restoration of health; and *educational gymnastics*, or exercise for the development of the healthy body. Although the exercises of one branch essentially differ from those belonging to the other, there are, however, many transitory forms of exercises which do not distinctly belong to either, but which make the two branches merge into each other. Medical gymnastics have been known in America under the name of "Swedish Movements," "Movement Cure," etc.; and we often find these names also bestowed upon the Swedish educational gymnastics by an ignorant public, in whose mind the relationship between the branches has made the two appear to be identical. Medical gymnastics not being within the scope of this treatise, we shall speak no more about them, except incidentally to mention their relationship to educational gymnastics.

Educational gymnastics have sometimes been given the prefix *hygienic* when their distinct aim has been more the preservation of health than the development of the physical powers. We also find many other names bestowed on this branch of gymnastics. Thus exercises that have a warlike tenor, such as fencing, boxing, wrestling, etc., have been called *military gymnastics;* such exercises as imply friendly competition, like the "tug-of-war," "putting the shot," etc., have been described as *gymnastic games;* and exercises whose chief aim is grace and beauty of form have been named *æsthetical gymnastics.* The word "gymnastics" can be applied to these exercises only if they are practised on gymnastic principles; and with this restriction they may also form a part of educational gymnastics, notably for advanced classes.

CHOICE OF EXERCISES.

The fact that gymnastic exercises should be systematic has led to many misconstructions of the word "system." For, claiming that a system must be "the complete body of an art or science," some persons have invented gymnastic "systems" which embrace every gymnastic possibility (such as exercises for the muscles of the face, tongue, eyes, and other "gymnastic foolery"); and in their hunt for completeness, they have finally ceased to be scientific, and have often practically been in direct opposition to their own theories and to ordinary common sense. A careful choice should be made, and the standard which decides the value of each exercise is the effect of the exercise on the body; *i. e.*, a system of gymnastics must be based on the laws of the organism. If founded on other considerations, it must disclaim all rights to being called a system, a science.

When speaking of "the development of the physical powers"

as the aim of gymnastics, we claim that, to have any value, this development must be harmonious, or there should exist an equilibrium between all the parts of the body, so that they form a well-balanced whole under the perfect control of the will — *mens sana in corpore sano*. A good posture is usually the expression of this development. Any partial development, such as we often see in athletes skilled in some particular gymnastic feat, has mostly been acquired at the expense of other parts, the result often being anything but a state of general health. We should practise gymnastics with health for aim, and not for the sake of excelling others in doing some particular exercise, — not for the sake of muscular development. Hence we say that each man is himself the unit by which his strength should be measured; for when all the parts of his body correspond to one another in relative size and strength according to the laws of nature, then he has attained the ideal of physical development, even if his muscular strength, as compared to that of others, is found to be comparatively small. In our attempt to reach this ideal, we not only have to encourage Nature in her normal activity, but we also have to prevent and overcome many tendencies to abnormal development. With this in mind, we have found a law that should govern our choice of exercises; for, now we say, that the greater or less value of a movement depends on its power to counteract and correct these tendencies, and that any exercise which encourages or produces such a tendency is not only valueless, but may also prove injurious; *i. e.*, the more a movement allows or produces faulty postures, the less is its value. It is the knowledge of these faults and of their means of correction that is the guide which adopts and rejects, compares and prefers. We find that the questions what shall be done in gymnastics, and how shall it be done, are really one and the same thing. This, however, should not be interpreted as if beauty of performance were to be the aim of gymnastics, but rather so that grace of execution is the final test that decides the perfection of the movement. For, not until a movement can be performed without faults — *i. e.*, with utmost beauty — has it had all its best effects on the body. Without a clear conception of faults in execution and of the greater or less influence of these faults, there can be no wise selection, — no reason or conscious motive for it. Ling has said that "the gymnastic value of an exercise depends upon how it combines the greatest effect on the body with simplicity and beauty of performance;" or, in other words, beauty of performance is essential to make an exercise effective.

PROGRESSION.

"System" is a name freely bestowed upon almost anything in the way of gymnastics, no matter how irregular the method be. If gymnastics are to be systematic exercise, the movements must follow each other according to some distinct and sensible rule: there must be progression. Not only should the exercises within each day's lesson follow each other systematically, but from lesson to lesson (from week to week) the movements should progress from the very gentlest or simplest, step by step to stronger or more complicated ones. It will be found that the best effect of an exercise is never received, if it is taken too early, — it may then even prove injurious, — and hence no exercise should be applied until the previous ones of the same kind have been thoroughly practised. Experience has proved that if some particular order is followed in each day's lesson, the exercises can be made more powerful, more complicated, and more advanced, without danger of over-exertion or other injurious results. Practical investigation has shown that to attain this object the following order is best:

1. Introductory exercises.
2. Arch-flexions.
3. Heaving-movements.
4. Balance-movements.
5 Shoulder-blade movements.
6. Abdominal exercises.
7. Lateral trunk-movements.
8. Slow leg-movements.
9. Jumping and vaulting.
10. Respiratory exercises.

As nothing has as yet been said about the nomenclature used in this manual, a short explanation will here be necessary.

The introductory exercises serve to gain general attention and muscular control, and to correct the base and general equilibrium, before anything of a stronger or more specific nature is done.

The arch-flexions consist of various forms of backward flexions of the trunk, given to expand the inferior portion of the chest, to straighten the dorsal part of the spine, and to cultivate the extensibility of the superior region of the abdomen.

The heaving-movements are various forms of lifting one's self by means of the arms on a horizontal bar or other apparatus, climbing ropes, ladders, etc. These exercises have for object to expand the superior portion of the chest, and to vault it upward, besides strengthening the arms. If they are not preceded by the arch-flexions, they are apt to produce a lameness in the upper region of the abdomen.

The heaving-movements increase the heart-beat, and should be followed by the balance-movements, as these consist of some easy leg-movements which require but little effort, and hence have the character of those movements which lessen the heart-beat, — the slow leg-movements. Besides, these exercises produce good and graceful balance, and consequently are needed also for this effect. If we recall that "the body should be a 'well-balanced' whole," we easily understand this.

The shoulder-blade movements are various forms of arm-movements which have the effect of correcting the position of the shoulder-blades, — an effect much needed by a majority of the growing generation.

The abdominal exercises are given to strengthen the abdominal walls, — a means of securing good digestion. These exercises are needed less by growing than by grown persons.

The lateral trunk-movements consist of various forms of bending and twisting the thorax sideways, etc. These movements strengthen the muscles around the waist, and affect internal organs contained in this region.

When we have proceeded thus far, we are probably again out of breath, and our heart-beat is much accelerated. Now the slow leg-movements are put in to lessen the heart-beat, and allow us to recuperate before we go on to more violent exercises. These leg-movements require next to no effort, whereas their effect in propelling the blood from the heart is quite forcible.

The jumping and vaulting can now be safely indulged in, as the previous movements have given us the intense muscular control necessary for these exercises, without at the same time producing any sense of fatigue. Jumping and vaulting improve the general elasticity of the body (and elasticity and grace are almost identical); besides, they produce courage, self-reliance, and a correct appreciation of space, etc.

These exercises are more or less violent, and invariably put us out of breath, and highly accelerate the heart-beat. They must necessarily be followed by some exercises to counteract these ill effects: such are the respiratory exercises. These consist of various easy movement of the arms (sometimes combined with movements of the legs or trunk) in rhythm with deep respiration, the respiratory act being the essential part of the exercise, the arm-movement being done merely to aid in expanding and contracting the chest.

When "tables of exercises" are made out, this order is usually followed as closely as circumstances will allow; and within each class the exercises follow each other in progression, step by step, from table to table. One class of exercises may also in a measure be dependent on the other, as far as its progression is concerned, — a fact to which we shall return later. To each table may be made a few additions, such as marching and running, introduced somewhere in the middle of the table, and another heaving-movement towards the end, etc.

The tables of exercises that are printed at the end of this book should be understood as being only instances of tables, not to be followed literally and indiscriminately. They are intended only as examples by which a teacher may model his own tables. For progression is not even, — every age and each sex has its distinctive predisposition in some respect, — inasmuch as physical strength does not increase in even pace with age, stature, or avoirdupois, and consequently this must be taken into consideration when the exercises are selected. For instance, jumping from a height is an easy thing for a child, whereas it is difficult for an adult. Women are capable of less speed than men, but more grace; their ankles, backs, chests, arms, and shoulders are weaker than those of men. They are very sensitive to backward flexions of the trunk, etc. Children have a very limber back, will do all arch-flexions easily;

whereas their ability to do the heaving-movements is very limited, etc. Hence it is essential to take individuality into consideration, so that each one may get what he needs, and yet so that the exercise is suited to the ability of all. If faults cannot be avoided, the exercise should be omitted till some future time, when the pupil is better able to execute it correctly.

Progression is usually made by increasing the speed of motion; the duration; the degree of force; by combining movements of different parts; by letting several movements immediately follow each other without pause; or by changing the rhythm of the movement. In women and children, we cannot demand so much speed, strength, or endurance as in men and boys of large growth; the necessary restriction is made by lessening the duration of the movements and yet giving the same number of exercises, or by giving a greater variety of exercises where we cannot safely increase the force of the movement. As a rule, for grown persons, progression takes the form of increase in *quality*, or force of movement; for children, in *quantity*, or number of movements. Girls before the age of puberty can safely have the same exercises as boys of the same age and strength.

APPARATUS.

The Swedish system of gymnastics contains free-standing exercises, and exercises on apparatus, yet it differs from other systems inasmuch as it is entirely independent of the apparatus, its movements usually being applicable to whatever may be at hand. This makes the system truly practical, as it is thus within reach of everybody, independent of wealth or other considerations. To reach good results by gymnastics, elaborate apparatus is not necessary; and, although we may be able to do more if we have apparatus of one kind or another, yet there are free-standing exercises that no apparatus in the world can ever substitute; and in gymnasium-work the free-standing movements should never be entirely omitted, but each lesson should contain at least as many of these as of the movements on apparatus. But, admitting that apparatus is useful, as it affords more possibilities, more variety, more recreation, there are certain qualities which we demand in good apparatus. 1. In the first place, it must be cheap,—a point that hardly needs argument. 2. In the next place, it should occupy but little room; for the floor of a gymnasium should be large enough to allow space for marching, running, free-standing movements, etc., and if the apparatus covers considerable surface, the room must be so much larger, it must have a separate "running-track," etc., all of which means additional expense. The apparatus used, if covering the floor-space, should be so constructed that it can be cleared away without a great loss of time: furthermore, apparatus made on these principles might be put into an ordinary room (schoolroom) without interfering with the use of that room for other purposes than gymnastics. This does away with the need of a gymnasium in schools that cannot afford one. 3. In order to be useful, the apparatus must also allow of a great variety of movements which are gymnastically unlike. So, for instance, a properly constructed horizontal bar can be used in exercises belonging to any one of the classes enumerated on page 5, whereas the average horizontal bar used in America is useful for but little outside of heaving-movements. The rowing-machine, bicycle-machine, and other "modern" apparatus are instances of apparatus which allow of only one movement, besides being expensive and covering a great deal of floor space. 4. The apparatus should also be so constructed as to allow a great number to use it at the same time. For, considering how few minutes a day are usually allotted to gymnastics, it is of great importance to see that no one is wasting time by having to wait until there is room for him on the apparatus. Thus a horizontal bar should be so arranged that several can use it at once, so that those who stand waiting have just time enough for a short rest. 5. The apparatus, when fulfilling the four previous conditions, is also suitable for all differences of age, strength, and sex, so that we do not need one machine for boys, another for girls, a third one for men, etc. But no apparatus — no matter how well constructed — is used to its best advantage, unless there be a skilled hand to direct it; and it is equally true that a good teacher will be able to do a great deal with whatever apparatus there is at hand — no matter how poor it be. When a teacher complains that he cannot do anything; that he cannot keep the majority of his pupils interested, etc., "because he has not the necessary apparatus," it is safe to say that it is *his* fault, and not the fault of the apparatus, nor of the system which he professes to teach; he simply passes judgment on himself as being unqualified to teach.

Below is given a description of the apparatus used in Swedish gymnasiums.

The Horizontal Bar (Fig. 1). — If properly constructed, this is the most useful of all the apparatus. Its dimensions are: length, twelve to eighteen feet; width, five to six inches; thickness, three to four inches. Its profile looks as in Fig. 1 *a;* the top is rounded so as to afford a good grip for the hands; the under side is grooved; this side is sometimes turned upward to allow of a different grasp or to better accommodate those who have small hands. The bar rests at each end on a bolt, stuck through a vertical beam in which the bar slides; this beam (one at each end), about ten feet high, is usually fastened to the wall, but may also be fastened free on the floor by some contrivance (as shown in the figure) so that it can be taken away if the space is needed. There are holes made for the bolts at intervals of two inches. If the bar has to be made shorter or longer in order to fit the room, its other dimensions must be changed accordingly. The bar should, however, never

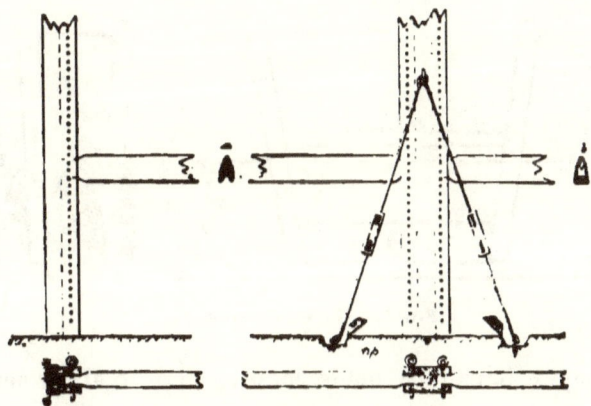

FIG. 1. — HORIZONTAL BAR.

be made too thick to afford an easy grasp, but its width may be changed to more than five or six inches so as to give sufficient strength. The bar is supplied with three or more wooden saddles (Fig. 2), which can be removed when not used. A bar twelve feet long will hold three of these saddles. A well-equipped gymnasium has a double set of bars to fit one above the other (double-bar) between the vertical beams; the other bar may have one plane side as shown in Fig. 1 *b* ; the bar, put low with this side up, will then serve for balance-movements.

The bars and saddles should be arranged so as not to rattle or shake when being used.

As compared to the round, narrow one, the Swedish bar has many points of superiority. This shape obliges the gymnast to use the muscles of his shoulders and chest more than those of his wrists when pulling himself up on bent arms; in that way this bar gives more chest-expansion than the other. In vaulting, the larger bar gives a firmer support, and hence is safer. The shape is an exact fit for the hands as they grasp the bar from above, thumbs behind, fingers in front. Besides, this bar

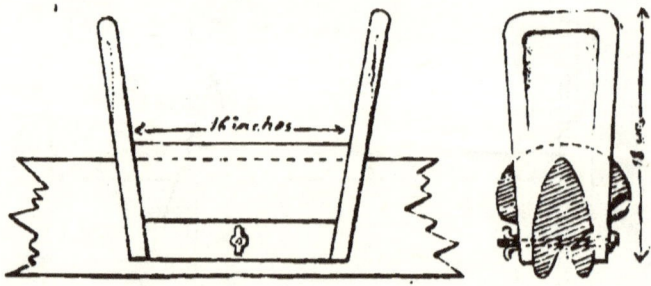

FIG. 2. — WOODEN SADDLES.

being stronger, it can be made long enough to allow several to use it at one time.

The Bar-Stalls (Fig. 3) constitute the next most important apparatus. They consist of vertical beams at intervals of three feet. These beams are about nine feet high, six inches wide, and two inches thick, and bars are fastened through them at intervals of five inches. The top bar springs out beyond the others, and the other bars are held in place by strips of wood screwed on to the front edges of the beams. The section of one of these bars is a rectangle one inch by two inches, with rounded corners. Wherever there is room, the walls may be lined with these bar-stalls. In front of the stalls are placed

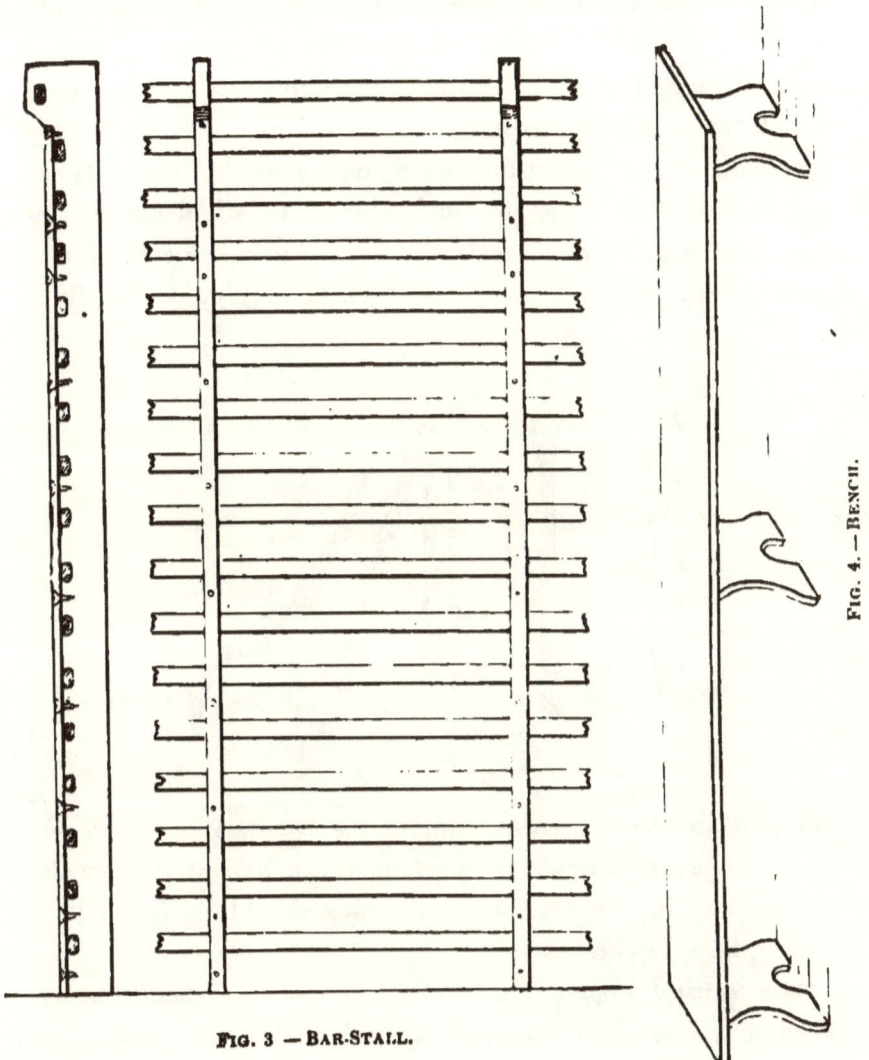

Fig. 3 — Bar-Stall.

Fig. 4. — Bench.

loose benches (Fig. 4), one foot high, the seats being nine feet long and one foot wide. There should be one bench to every three bar-stalls.

The Poles (Fig. 5 a) are round wooden spars with a diameter of two and a half inches. They are fastened to the ceiling by means of hooks, and they reach nearly to the floor. When not in use, their lower free ends should be tilted against the wall, and held there by means of a loop or other contrivance, so that they are out of the way. There may conveniently be

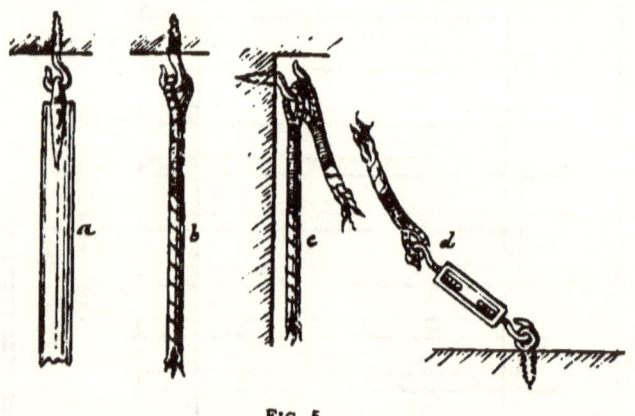

FIG. 5.

one pole in front of each beam of the bar-stalls, which is suspended so as to hang about one foot and a half away from the beam when in use. Six or more poles should be found in a well-equipped gymnasium.

The Vertical Ropes (Fig. 5 b) are smooth (without knots), have a diameter of one and a half inches, are suspended from the ceiling, and come within a few inches of the floor; their lower ends are covered with leather. Six or more of these may be put in a row parallel with the horizontal bars, and a few feet away from them. The ropes, as well as the poles, should be at least one and a half feet apart; when not in use, they are gathered by a loop running through a pulley, and hoisted out of the way.

The **Horizontally Inclined Rope** is fastened at one end by a hook in the wall close to the ceiling (or in the ceiling close to the wall); its other end is fastened to the floor near the opposite wall (or in the wall near the floor). This end is so arranged (Fig. 5 d) by means of hooks and screws that the rope can be easily fastened and firmly stretched when in use; when not in use, this end is unhooked, and the rope gathered to the wall on which its upper end is fastened. The rope should be so arranged that the incline is 45° or less; and the whole length of the rope should be from forty to fifty feet. From the hook at the ceiling a vertical rope may be suspended so that no time is wasted by the pupils having to climb back the same way they came, but one rope serves as a "feeder" for the other. If space allows, there should be two inclined ropes in a gymnasium.

Wooden Ladders. — These are very useful pieces of apparatus, especially for children. They can be constructed in many different ways, one of which is shown in Fig. 6. This (the vertical) ladder is suspended from the ceiling, and reaches to about two feet above the floor. The hooks from which it hangs are fastened to the ceiling at least two feet from the wall. The lower end of the ladder is allowed to swing free, or it is fastened to the wall by means of hooks and "screw-eyes," according to the purposes for which the ladder is used at the time. Another (the horizontal) ladder may be arranged so as to hang with the long side up (Fig. 7), its lower edge being about six feet from the floor. This ladder is best fastened by ropes and pulleys so that it can be raised or lowered. One

16 EDUCATIONAL GYMNASTICS

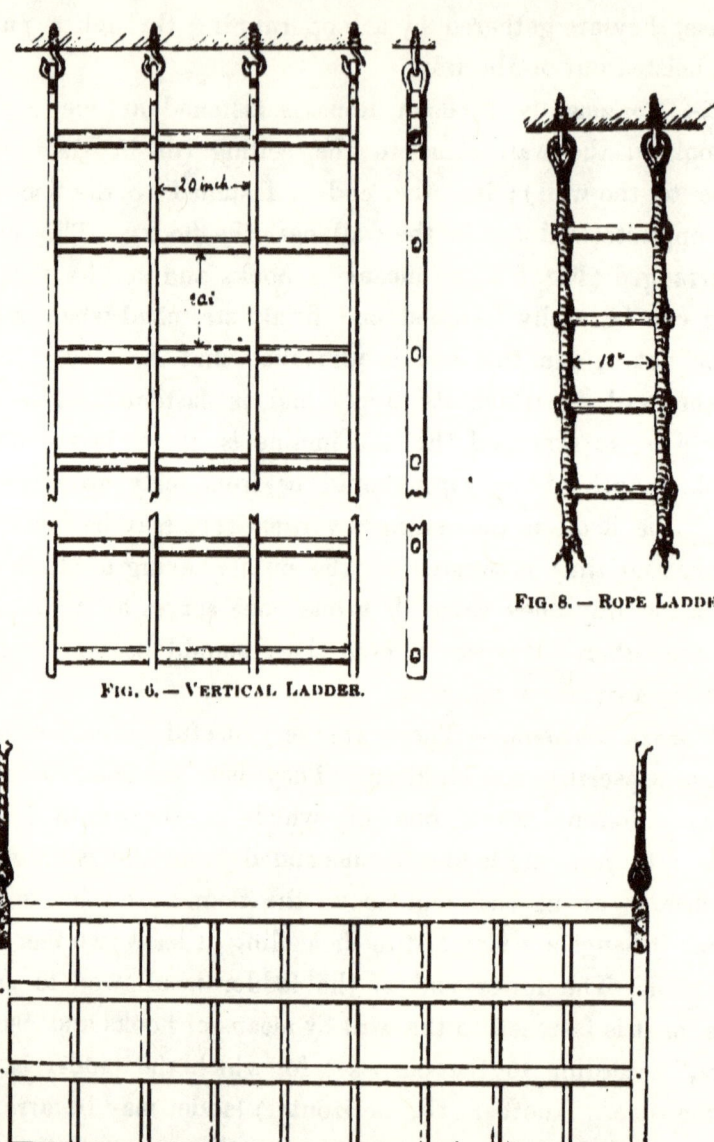

Fig. 6.— Vertical Ladder.

Fig. 8.— Rope Ladder.

Fig. 7.— Horizontal Ladder.

ladder of each kind is enough in an ordinary gymnasium. If space allows, there may also be hung six or eight rope ladders (Fig. 8), whose lower ends swing free when the ladders are in use. At other times they are hoisted out of the way like the vertical ropes.

The Vaulting-Horse.— The modern horse has the appearance of Fig. 9. The body is about six feet long and one and a half to two feet broad over the hind-legs, and six to eight inches narrower over the front-legs. It is made of light wood strengthened by iron bands, and it is wholly covered by leather.

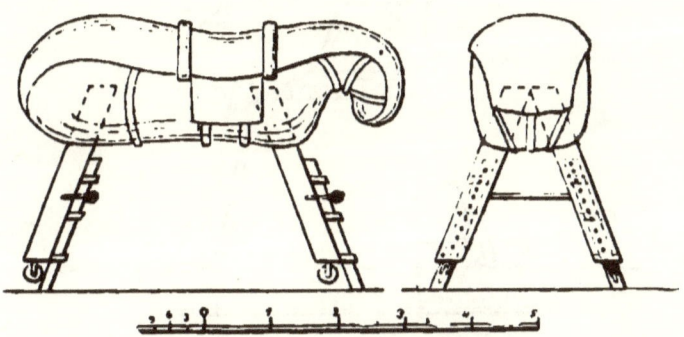

FIG. 9.— THE VAULTING-HORSE.

Along the top is a long piece of roughly finished leather forming a blanket and a saddle with low pommels, held on by straps buckled under the body. The legs are square, made of hard wood, and firmly fastened into the body, slanting outward so as to make the horse steady. They are supplied with casters so as to make it easily movable, and on their front sides are fastened loose square legs running through iron frames. These legs are movable so that the horse may be made higher or lower, and they are held in their places by bolts running through holes made through both legs. When the horse is to be moved, these legs are raised entirely above the floor. Although much more expensive than the old-fashioned horse (Fig. 10), it is far more practical than that one. However, the latter is a very good piece of apparatus, and will do almost as well for those who cannot afford the modern one.

The Vaulting-Box (Fig. 11). — This is a wooden box made in compartments. It has slanting sides and a top upholstered in leather. All the corners and edges are rounded. The top measures about one and a half by five feet, and the bottom three by five and a half feet; and with all the compartments,

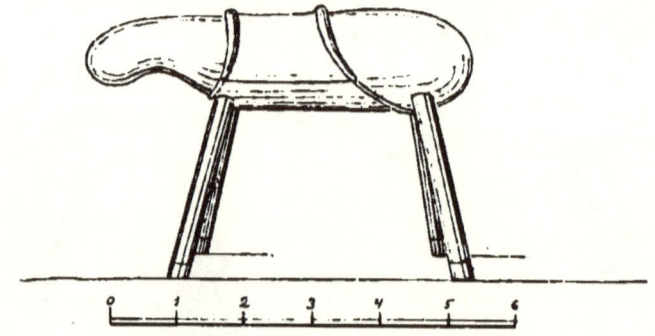

FIG. 10. — OLD-FASHIONED HORSE.

three or more in number, it stands from four to five feet high. The compartments are supplied with handles or holes cut in the wood, so that they can be easily lifted off. This piece of apparatus is comparatively cheap, and will in a measure substitute the horse; besides, it allows of certain movements (somersaults) which cannot be as safely done on the horse. A complete gymnasium has one horse and one or two of these boxes.

In a gymnasium there should also be found two or more loose ropes about one inch in diameter and about nine feet long. These are used for free jumping and for various other purposes; when used for jumping they are usually put into stands especially made for this purpose, as found in most gymnasiums; but they may be fastened into the stall-bars or some other place quite as well.

The horizontal bar, especially if a double one with loose saddles, will form a substitute for any of the above enumerated apparatus; thus if you can afford only one piece of apparatus, this bar should be your choice. If you are called upon to apply the Swedish system in a gymnasium already fitted up,

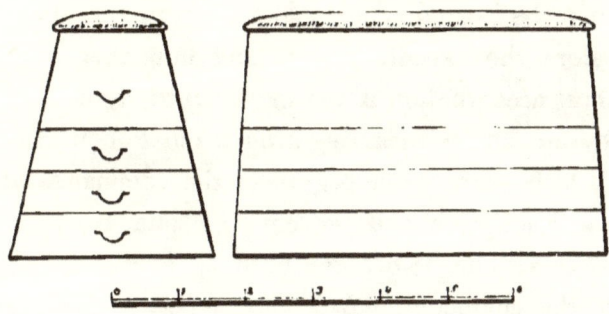

FIG. 11. — VAULTING-BOX.

you will need no new apparatus, but can use whatever there is at hand; for, the exercises, and not the apparatus, make the system.

WORDS OF COMMAND.

A lesson in gymnastics is best conducted with the class arranged in ranks, military fashion; this cannot be strictly carried out when the apparatus has to be used; but, whether the exercises are free-standing or on apparatus, nothing should be done unless the teacher has ordered it.

For this reason he uses words of command. These consist

of two parts: one part, the *preparatory*, describing the exercise — it tells what should be done; the other, the *executory*, gives the time when it is to be done, and it also indicates the speed of motion. The preparatory part may sometimes be preceded by another, *cautionary* part, telling the pupils what to avoid — what not to do. This is often used to warn against common faults in gymnastic movements. For instance: "*Without moving the shoulders* (cautionary), *head backward* (preparatory) — *bend!* (executory)." The first two parts are pronounced with loud, distinct, and moderately slow voice, so that every one can fully understand what is to be done and be fully ready for it when the executory command is given. A short pause is made before the executory part, and now this is given with sharp, short accentuation if the movement is to be performed in rapid rhythm, and with a lingering intonation if the speed of motion is to be slow. For beginners the commands are longer and more descriptive, and the counts, "one, two, three, four," or "one, two; one, two," are used as executory parts. For instance, the command, "*Left foot sideways place and arms upward — strétch!*" when given to beginners would be: "*Left foot sideways place — óne!*" and after that has been done: "*Arms upward stretch — óne! Twó!*" and so on. For some apparatus-movements, the pupils take the commencing position at the command, "*Reády!*" after the position has first been described. The movement is then executed to commands as usual. The commands throughout this book are printed in Italics with quotation-marks, and an accent indicates whether the executory word is to be short or prolonged according to the desired speed of motion.

This method of leading a lesson has been greatly objected to by various persons. Some have the idea that a better result is obtained if the teacher stands in front of the class and performs, while the class imitate his movements; but you will find that in this case the teacher usually is the only one who executes each movement as it ought to be. The pupils have their attention more concentrated upon watching and imitating the teacher than upon the movements they themselves are doing, and the consequence is that no movement is done to its fullest extent, and the best effects are lost. If the movement is to take place for the first time, and the teacher has reason to suppose that the command does not sufficiently describe it, then he may show the movement before commanding it. If he himself, for some reason, does not wish to perform before the class, he selects a skilled pupil to show the movement, and then proceeds to command it, after the pupil has taken his place in the class. The last arrangement is often the best, as the pupil is liable to get into a faulty posture, and the teacher then has the opportunity to correct it and thus show the class what to avoid. Whoever is showing a movement should take care *to face the same way as the class;* for left and right will otherwise easily be mixed by dull pupils.

When the movement is done, the class stands still in the new position, until something else is ordered. And now the teacher occupies the pause by moving about, correcting by word or action those who have a faulty posture. All these corrections should be made very quickly, so that the class is not kept in a tiresome posture for more than a brief period, or their attention will soon be lost. As a rule, the manual corrections — a pull or often the slightest touch is all that is necessary — will be found to be quickest in result and consequently the most practical; and usually there is time to correct only two or three pupils.

Repeat the movement if necessary, but do not keep the class standing in one position for any length of time.

Now if the teacher were performing with the pupils, how could he make these corrections without temporarily interrupting the lesson? To keep the strict attention of the class, he would have to overlook the faults, and thus neglect half his duty; for a teacher's duty is not only to instruct what to do, but it is equally to correct and prevent that which should not be done.

Some advocate exercising to music as being the best way of practising gymnastics. Here the pupil thinks more of the rhythm in which to do the movement, than of the manner in which it should be done; and, his attention being divided, one of the objects of it must suffer: he sacrifices the effect of the movement for the sake of keeping time. Furthermore, there are very few *gymnastic* movements that are rhythmical; in fact, only walking and some other oscillatory movements can be classified as such, whereas every gymnastic movement has a rhythm of its own distinctly different from the rhythm of music. If you desire to do a movement rhythmically, as may sometimes happen, your commands, "*one! Two! One! Two!*" ... or "*one! Two! Three! One!*" ... will furnish the necessary "music." In a discussion on this point, a prominent educator of this city said that gymnastics to words of command were not mechanical enough to be recreative, whereas gymnastics to music furnished this recreation, especially for those whose occupation is brain-work. To this we answer, that in order to reap all the benefits of recreation, you must put your mind into it; whereas, if you play in a mechanical way, while your mind is busy on something else, your recreation will do anything but give the required rest. What is it a brainworker most needs? He needs to put his mind, which is usually bent on something outside of him, into his own body; he needs to take exercise where every movement is done with full attention, full volition; and it is only then that he will reap all the benefits which can be received from exercise. If you insist upon using music, the best form of movement would be dancing, which exercise is both rhythmical and recreative.

Again, others recommend series of movements which the pupils learn to perform in a certain unvarying order. In this kind of exercise the pupil usually pays more attention to the order to be followed — "what comes next?" — than to the movement he is doing at the time; and, his attention being divided, he will either forget the order of the series, or neglect the correct execution of the movements.

Words of command necessitate discipline: the class must obey when the teacher orders. As for this arrangement being "tiresome for the pupils," "too much military style for ordinary citizens," "too pedantic," etc., we wish to state that this is the least tiresome method of applying gymnastics. Moreover, discipline should be taught not only to military men but to everybody, for no one will have the same power of self-control unless some time in his life he has had to obey; and this discipline is by no means any encroachment upon the pupil's "rights as a free citizen of a free country." Only those who know what restriction means can truly appreciate liberty, and make a good use of it. This discipline need not be pedantic, if the teacher knows how to let rest and activity, play and work, alternate. Frequent changes between these will make activity more earnest, rest more playful; and they will teach the pupils how to better distinguish between active and passive rest. In the education of children, we all know how a teacher without discipline is unable to impart knowledge to his pupils; and this is just as true with reference to gymnastics. Besides, children all like a good disciplinarian, who knows how to combine firmness with good humor and a kind disposition; they willingly obey, and he will accomplish more with them than any one else.

Many claim that these soldier-like Swedish exercises are not suitable for girls, because "they give to our women a military gait and posture." The soldier stands and walks erect, and moves with precision, not because he is a soldier, but because he has learned how to carry himself; and it certainly does not lessen the beauty of the fair sex, if they also learn how to stand and move in best conformity with the laws of nature. Indeed, the saying that "man was made to walk erect," refers to man and woman equally; and if woman is to be man's equal, she should bear in mind that "the only true basis for culture is the physical basis;" for, a sound mind requires for existence that it should dwell in a sound body; and a system of gymnastics, founded on the laws of nature, is suitable for women as well as for men, whether it is taught to words of command or not.

Every lesson in gymnastics should begin with the command, "*Atténtion!*" which word implies that the pupils should stand still (wherever they may be) in an active and careful posture, ready for work. When a rest is desirable, the command, "*Stand — at ease!*" or "*In place — rêst!*" indicates that the pupils may take a restful posture without leaving the ranks, or the place where they are standing at the time. To omit either of these commands, where they should be given, is to allow the pupils to become negligent. Whenever a new exercise is shown to the class, it is best to let them stand at ease; and, if they show signs of fatigue or lessened attention, command " *In place — rést!* " and let this be soon followed by " *Atténtion!* " In that way, whatever is done is done by your will and not by that of the pupils, — or at least they will think so! Under all circumstances, as long as the lesson lasts, the teacher should have absolute authority in the gymnasium, whether the class be at work or at rest.

GYMNASTIC DRESS.

There is a common supposition that gymnastics require a special dress; and, on the other hand, a great many claim that you should wear the same dress when you exercise that you do in ordinary, every-day life. In gymnastics it is essential that the dress be loose so as to allow full freedom of motion; consequently collars and anything tight around the limbs should be removed. As for corsets, it is to be hoped that no one will be rash enough to practise gymnastics while embraced by this enemy of womanly health and beauty. The costume should be light in weight, so that the increase of bodily heat may not become excessive: however, it need not be any thinner than what should commonly be worn in-doors, for such a dress, especially in the winter, should always be light if we wish to avoid taking colds. For school-children there will be no need of special gymnastic dress, when the exercises are practised in the schoolroom; for, in this case the movements are necessarily simple, and have none of the violent nature which would require a still looser dress for more freedom of motion, or a lighter one to prevent the great rise of temperature that is the usual result of violent and powerful gymnastic movements.

If the girls are dressed in a common-sense way, *i.e.*, in short skirts that do not drag heavily on the hips, with a blouse or other loose waist and without corsets, they will be able to partake of the gymnastic exercises in the schoolroom. If gymnastics are made compulsory in the schools, girls will have to dress accordingly, and perhaps in that way the much-talked-of "dress-reform" can be brought about. For, the average girl, thus accustomed to perfect freedom of motion, would probably not be willing to give up this liberty for the sake of wearing the conventional stays, for which she would have no more use, since her muscles would then be developed to do the work for which nature has designed them, but which a mistaken idea of beauty has transferred to steels and whalebones.

As for boys, they are always dressed for gymnastics; and if they remove their coats, and possibly their vests also, no other change need be made, not even in the gymnasium.

In a gymnasium, however, where more time is allowed for gymnastics, and where there are usually separate dressing-rooms, it is desirable that a regulation costume be worn, especially by grown women; for, a woman's dress of to-day is thoroughly unfit to be used for gymnastic purposes, as it admits neither of free respiration nor of any other freedom of motion. The best gymnastic dress for women is the divided skirt with a blouse-waist (Fig. 12). The skirts are fastened above the knee, and fold over, reaching nearly to the ankle. One form of dress (Fig. 13) consists of knee-breeches, short skirt, and blouse-waist. This, however, is less practical than the other, besides giving a more awkward appearance. Those who object to wearing gymnastic dress may wear a very light skirt of usual length; but they will soon conform to the regulation dress on account of the greater freedom of motion which it allows.

The shoes should be light, the best kind — for men as well as for women — being low shoes with cloth tops, rubber soles, and low heels (or none at all). The rubber sole has the advantage over the leather that it prevents slipping, and thus averts many an accident. Low shoes are better than high ones, because they enforce a stronger activity of the ankles, and thus develop and strengthen them.

The best gymnastic dress for men consists of shoes as above, moderately wide trousers of flannel or duck, a loose shirt of the same material with turned-down collar, and a sailor's belt. The collar and tie should not be so large that they "float before the wind," as they will then be in the way. The whole dress should be white, with possibly a narrow ribbon around the collar and sleeve-linings. Fancy costumes belong to the circus, and are not becoming in a gymnasium. The low-necked and sleeveless shirt is hardly decent in any other place than one's own

FIG. 12.—DIVIDED SKIRT, BLOUSE WAIST.

FIG. 13.—OLD-FASHIONED GYMNASTIC DRESS FOR WOMEN.

room when no one is present; besides, the naked neck and arms are liable to get chafed from the apparatus.

For ordinary gymnasium work it will be sufficient for a man to remove his collar, cuffs, coat, vest, and suspenders, and to change his boots, as this takes but little time. His gymnastic dress he may save for exhibitions, to give a more harmonious look to the class work.

PART II

MOVEMENTS OF THE SWEDISH SYSTEM OF GYMNASTICS

DEFINITIONS AND GENERAL THEORIES.

"THE oneness of the human organism, the harmony between mind and body," constitutes the fundamental principle of Ling's gymnastics. This harmony is preserved and developed in the healthy by gymnastic movements, aided by hygienic agencies, such as fresh air, food, drink, sleep, dress, etc.

A gymnastic movement is a movement that has a definite space, time, and force. "Stretching a limb or making frictions along a muscle, nerve, or vessel, in a direction and manner and with a velocity and force all previously determined, and for a distinct purpose, are instances of gymnastic movements."

The limits of a movement are determined by the laws of gravitation and by the sphere of activity of the muscles. If the movement is not in conformity with the laws of the organism, it not only loses its value, but it may also prove injurious. Hence the anatomical structure and the physiological activity of the body, as related to motion, should be familiar to those who apply or superintend gymnastics.

A movement is a change from one position to another,—a chain of intermediate positions; and it is rendered exact when the commencing position, terminating position, the line of motion, and the rhythm of action are clearly and severally determined. An accurate movement can be done only from a correct commencing position; and if it is correctly executed, the terminating position should be as correct: hence the position that a movement creates is the surest means by which we can judge of the value of the movement. A point to be remembered is, that when a movement is *disfigured*, it generally does more harm than good; or, a movement has the best effect on the organism when it is performed in a graceful or beautiful manner; or, to quote Ling, " a correct position is the first condition for the utmost beauty and power of action,— for completeness and decision."

If we remember that every muscular effort is involuntarily preceded by an increased inhalation, and also that a contracting muscle absorbs about double the amount of oxygen that it does while at rest, we easily understand why free respiration is absolutely necessary in all exercise, and also why the greatest importance must be laid upon the development of the respiratory organs. If a position or a movement interferes with free respiration, it must be rejected. With reference to this point, Hj. Ling says that " every correct active movement, executed with undivided attention and intense will-power, is to be considered in a measure as a respiratory movement. This condition is expressed to every experienced eye by the way the head, neck, shoulders, chest, etc., are carried, and is lost by so slight a thing as a fall of the eyelids, etc. This is true of all positions and movements, from the simplest to the most complicated. . . . Hence all execution depends not only upon the quantity, but also upon the quality, even in jumping. . . . There is no greater mistake than to suppose that there could be any one active position deviating from this simple rule. . . . Hence the necessity of making a distinct difference between standing, sitting, kneeling, passive positions, and the corresponding active ones: the former express rest or lack of attention, the latter activity and readiness for work." The little word "*Breathe!*" should be frequently used during a lesson in gymnastics, so as to make sure that no one is forgetting to comply with this important law of nature.

According to the point of view from which they are considered, movements may be classified in widely different ways. Physiologically, movements are: 1. Voluntary, — those that are the condition of meditated action. 2. Involuntary, — those over which the will has no control, such as the heart's action, the movements of the stomach and intestines, etc. Or, movements may be spoken of as simple and complicated (compound, complex), etc. All these classifications belong to the general kinesiology (science of movements), and hence need no mention here. Gymnastically, movements are active, passive, resistive, and assistive. The last three classes belong to medical gymnastics, and will not be described here, even though some of their forms may be borrowed and used in educational gymnastics. An active movement is a minutely determined, voluntary, muscular movement. In gymnastics it must have a distinct gymnastic purpose, in contrast with other voluntary movements.

Physiology teaches us that every muscular movement should be considered as having three phases: 1. The nerve and muscle are getting ready to work, and yet there is no visible movement. 2. The muscle contracts. 3. The muscle relaxes, or returns to its former length ("medium position"). For this reason a movement begins and ends slowly and with but little force, whereas during the intermediate period the speed may be increased, as the force grows. As a rule, it may be said that it is easier to do a movement quickly or with moderate speed, than it is to do the same movement slowly. In this way, one and the same movement repeated may serve as its own progression, simply if the speed of motion is changed. This, however, does not refer to *all* gymnastic movements, for there are movements which should always retain all their original speed in order to keep their gymnastic character and effect; and, at the same time, there are others which are to be executed slowly, even when first used. This will be better understood through the description of movements which follows.

For ease of description, a great many authors have classified movements, as head-movements, arm-movements, leg-movements, etc. In this book the exercises will be classified as on page 5, with the difference that after the introductions will be placed leg-movements, which class will contain introductory leg-movements as well as balance-movements and slow leg-movements, as this arrangement will somewhat simplify the descriptions and thus save space.

Each movement will be supplied with its command. To abbreviate the latter, it will be written like the following example: "*Left (r.) foot forward place and arms upward — stretch! Trunk to the left (r.) — bend!*" this meaning that when the left foot is placed forward, the body should be bent to the left, whereas if the right foot is forward, the body is bent to the right. The movement should be repeated at least three times to one side, before the commencing position is changed and it is executed to the other side.

To shorten the nomenclature, the following abbreviations have been used.[1] The capitals denote the various parts of the body; whereas small letters, or a capital followed by small letters, indicate the position and movement. Thus: —

[1] See page 273 for complete list.

A. = arm.
B. = back.
F. = foot.
H. = head.
K. = knee (Kn.).
L. = leg.
T. = trunk.
abd. = abduction.
add. = adduction.
circ. = circumduction.
backw. = backward.
cr. = crook (refers to leg).
elev. = elevation.
ext. = extension.
fall. = the trunk bent backw.
flex. = flexion.
flg. or fling. = flinging.
forw. = forward.

gr. = grasp.
hang. = hanging.
hlf. = half (when only one leg or arm is used).
hor. = horizontal.
l. = left.
ly. = lying.
pos. = position.
r. = right.
rot. = rotation.
sidew. = sideways.
sitt. = sitting.
st. = standing.
std. = stride (refers to the feet).
stp. = stoop (trunk bent forward).
str. = stretch (refers to the arms).
swim. = swimming.
turn. = the trunk rotated.
yd. = yard (refers to the arms), etc.

The nomenclature may at first seem odd, but the teacher will soon become familiar with it and find it a great help.

To save repetition in the descriptions, the exercises within each class are grouped together according to commencing positions, wherever this can be done. Thus the descriptions do not follow any progression; but after each class is given a list, in which the exercises are enumerated in progressive order, the list merely giving their names. These progressive lists do not claim to be correct in every particular, but they may prove a valuable help to the teacher, when he has to make out his own tables of exercises.

INTRODUCTORY EXERCISES.

Introductions are exercises used at the beginning of each lesson to gain general muscular control, to correct the base and balance, and to improve the attention. At first only simple leg-movements are used, such as taking the various walk-standing positions, and changing feet, etc. As the class advances, we add one or more exercises to these, so that, for an advanced class, a complete table of exercises should contain before the arch-flexions a respiratory movement, leg-movements, arm-movements, a lateral trunk-movement, and exercises for order and rhythm. These movements are chosen from among those which have been done at some previous lesson, so that they now require but little effort, — movements which we wish to repeat so as not to forget the positions which they produce, etc. Some exercises, which do not have a sufficiently distinct effect to belong to any particular class, but which are necessary as preparations for other and similar movements, must be classified as introductions. The first two or three lessons for a class which has had no previous gymnastic training are largely made up of this kind of exercises. On the other hand, there are movements which, although they are introductions to others of the same kind, yet have a distinct effect of their own; and on that account these exercises are not only introductions, but also what we call real gymnastic movements. Consequently, all exercises of this kind will be described in the various classes where they belong, and only such movements as can have no other place will be included under the heading "Introductions."

Fundamental Position (Fig. 14). — The first thing to teach pupils is to take the fundamental standing position (fund. st. pos.). This corresponds to the position of the soldier, except that the feet are at right angles. Thus the position is as follows: heels together and on the same line; feet turned out equally, and making with each other an angle of 90°; knees straight, without being stiff; the body erect on the hips, which are drawn slightly backward; the chest well expanded; the shoulders drawn backward and downward so that they are level; the arms hanging down of their own weight close to the body, with the hands extended, palms resting on the thighs "behind the seams of the trousers;" the head erect with the chin drawn in; the eyes looking straight forward, not downward; the weight of the body carried by the balls of the feet.

The angle of 90° gives a larger base than that of 60°, and has the advantage, gymnastically, of making it easier to find the direction "sideways forward," this now being in direction of the foot, which is pointed midway between sideways and forward.

If a person be knock-kneed, or have very large calves, he will not be able to stand with his heels together, but may then be allowed to have them a little apart.

To make the pupils take this position, give the command, "*Atténtion!*" or, if they already are attentive, but standing in some other position, and you wish them quickly to take the fundamental position, the command is, "*Position!*" which means, "Take the fundamental position in the quickest possible manner."

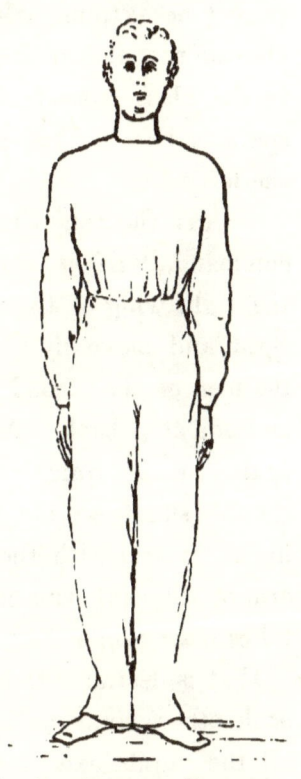

FIG. 14.—STANDING POSITION.

When a rest is desirable, you command, "*In place — rést!*" which allows the pupils to take easy positions, and talk, provided one heel is always kept in place. If you wish them to stand in easy positions without talking, you command, "*Stand — at ease!*"

How to Arrange the Class. — Before the lesson (in a gymnasium) begins, at the command, "*Fall in ranks!*" the class forms in two ranks, one behind the other. Arrange the class so that the tallest one stands as first man on the extreme right (right wing) in the first (front) rank, the next one in size behind him as first man of the second (rear) rank; the next two tallest are placed in the same way on the left wing (the extreme left); and this way you arrange the whole class so that the smallest ones stand in the middle of each rank. If the class consists of an odd number of pupils, an empty space is left in the second (rear) rank next to its last man on the left wing.

To get the ranks in a straight line, and to correct distances, command, "*Right — dréss!*" All, except the first man on the right wing of the first rank, turn their heads quickly to the right, and move by little steps so that they are all in a line, the men of the second rank standing behind those of the first and one step back. At the command "*Frónt!*" the heads are again turned quickly forward. The men of the first rank should stand so far apart that, if each one slightly raises his right arm with the elbow a little bent, it touches the left arm of the next man on his right; this adjustment of distances takes place while the " right — dress " is being executed.

If it is desirable to have the pupils at a greater distance from each other, command, "*Whole distance, right — dréss!*" when all the pupils except the first man on the right wing of each rank march to the left flank; each one halts as soon as he is so far away that he can just touch the left shoulder of the pupil on his right side with the fingers of his outstretched right arm. The pupils keep their right arms lifted and their heads turned to the right, until the teacher commands, "*Frónt!*" If a still greater distance is desired, the teacher commands, "*Double distance, right — dréss!*" This is done on the same principles, except that the pupils stretch both arms sideways, and halt so far apart that their fingers touch.

Open Orders. — The ranks are opened at the command, "*Open ranks — márch!*" The second rank takes two steps backward, beginning with the left foot, and halts (count "*One, two, three!*" if you so wish), taking care to form an even line.

To open orders, command first, "*Count — twos!*" Within each rank the first man on the right says "*One*," while he turns his head to the left; upon hearing this the second one turns his head to the left and says "*Two*," the third one says "*One*," the fourth one says "*Two*," and so on all through the rank. If the ranks are closed, only the first rank counts, the men of the second rank taking the same numbers as those who stand in front of them. The next command is, "*Numbers two (one), one step backward — márch!*" The designated numbers take one step backward with the left foot, and bring the right foot down beside the left one. For a little practice, you may now command, "*Numbers one and two, change ranks — márch!*" The twos take one step forward, the ones one step backward, etc.

To close orders, command, "*Numbers two (one), one step forward — márch!*" and then, "*Close ranks — márch!*" The second rank now takes two steps forward, beginning with the left foot, and halts. For advanced classes, the shortest command is, "*Close orders — márch!*" when numbers two (one) step forward into their places in the respective ranks, and the ranks then close as before.

Another and favorite arrangement is to count threes instead of twos, and, after having opened ranks, to make numbers one take one step forward and numbers three one step backward. This arrangement affords the teacher the best survey of the class.

If the class is facing the flank, at the command, "*Open ranks — march!*" each rank takes one step to its respective side, so that the ranks separate. To open orders, command, "*Numbers two, one step to the right (l.) — march!*" or, "*Numbers one, one step to the right (l.); numbers three, one step to the left (r.) — march!*" etc.

If space does not admit of the whole class in one division, it may be formed in two divisions, one behind the other and six steps away from its second rank. The commands will then be the same as above. If there are less than twelve pupils in a class, they are best arranged in one rank.

Facings. — To face to the flank, command, "*Right (l.) — face!*" The pupil turns on his right (l.) heel 90° to the right (l.), and supports the motion by the ball of his left (r.) foot; when this is done, he places his left (r.) foot beside the right (l.) one in fundamental position. Facing to the right is done on the right heel, facing to the left on the left. This is directly opposed to American military drill, where the turning is always done on the left heel, but it is true to the gymnastic principle which teaches that movements should be done equally on both sides. Half-facing is executed on the same principles, to the command, "*Right (l.) oblique — face!*"

To face the rear, command, "*About — face!*" On the left heel turn 180° to the left, supporting the motion by the ball of the right foot. When the semicircle is completed, the right foot is placed beside the left one in fundamental position. This also differs from the military drill, where the facing to the rear is a most complicated affair. The Swedish way of turning always to the left is intended to make us as well balanced on the left as we naturally are on the right; besides, the motion is executed on exactly the same principles as right and left face, which greatly simplifies matters. Not to make us one-sided, however, we should occasionally practise "about — face" to the right, the command being, "*Right about — fáce !*" The motion is then executed on the right heel, and supported by the ball of the left foot. An occasional "right about — face," is a good means of wakening a sleepy class, or of refreshing the attention a little.

In all facings, the arms should be kept close to the body and the shoulders level.

During a lesson in gymnastics, the class should usually be facing the flank.

Marching. — Although marching does not strictly belong here, we will, however, describe it in connection with the above movements, since it forms part of the standard "drill." We totally discard the "balance-step," and similar exercises, as being too antique for modern drill. To start the class, command, "*Class, forward — márch !*" when every pupil immediately begins marching by bringing the left foot forward first, the right one following with a speed of a hundred and fourteen to a hundred and sixteen steps a minute; the length of each step should (for adults) be about two feet and a half. In marching, the chest should be carried well forward and the head erect. The shoulders should be kept level, and to prevent them from swinging forward and backward, the arms should oscillate in even rhythm with the legs by a slight flexion and extension of the forearm, the limbs of opposite sides moving simultaneously. As the foot touches the ground, knee and instep should be extended so that the ball of the foot and the heel touch the ground simultaneously. If the heel strikes first, the step loses its elasticity.

To increase the length of the steps, command, "*Long steps*

— *márch!*" when the steps are increased to about three feet. To regain the ordinary length, command, "*Ordinary — márch!*" when the steps are again shortened. To shorten the steps to about a foot and a half, command, "*Short steps — márch!*" and again, "*Ordinary — márch!*" The short steps should be a little more marked than ordinary marching. If you wish to keep the class marching in the place where they stand, command, "*In place — márch!*" (or, "*Mark time — márch!*"), when the pupils slightly lift their feet, "and make a semblance of marching, without gaining ground." If from that you wish to get ordinary march, command, "*Forward — márch!*"

Marching backward is done to the command, "*Backward — márch!*" and resembles "forward — march," except that the steps are now only half as long as in the latter.

If you wish the class to change step, command, "*Change step — márch!*" The foot that is behind is placed beside the front one, instead of passing it, and the latter takes a new step forward. A continuous change of step ("*With change of steps, class forward — márch!*"), a movement somewhat resembling polka, is an excellent means of quieting down the rhythm of marching, if it has grown too rapid. Besides, this form of semi-dance is excellent for developing a graceful and light step. Care, however, should be taken that the step does not become too elastic, as is often apt to be the case. On the other hand, the short step usually quickens the rhythm.

If the class does not keep time, command, "*Mark every three (five) steps — márch!*" At every third step the foot is put down with an increased pressure, knee and instep being stretched as usual. The pupils should not be allowed to slam the foot down with bent knee, trying to make all possible noise. Always make the marking occur on odd steps, so that it will be done by each foot alternately.

Running takes place at the command, "*Double quick — márch!*" when the whole class simultaneously come forward with a speed of about a hundred and sixty-five steps a minute. In running, the body should be gently inclined forward, the head thrown back, and the mouth closed. The forearms should be lifted to horizontal position, with firmly closed hands, the elbows well backward and close to the body. The forearm is moved in a straight line slightly forward and backward in rhythm with the legs. The steadier this motion is, the easier is it to breathe naturally and deeply. A distinguished Swedish writer has described walking as being "a constant falling forward, where the weight of the body is received by each leg alternately;" and this is still more true of running. If you once know how to run with the least expenditure of energy, this form of motion is no more fatiguing than walking, — for short distances at least.

Running becomes more elastic and less exhausting if done on tip-toe. Command, "*On tip-toe — márch!*" . . . "*Double quick — márch!*" To regain ordinary march from running, command, "*Common time — márch!*"

To Halt. — If the class is marching, and you wish them to stop, command, "*Class — hált!*" The class take one more step, and bring the backward foot down beside the other one (thus you count, "one, two"). If the marching is done on tip-toe, the halt is done on tip-toe, and the heels are then lowered (count, "one, two, three"). If halt is commanded from double-quick, the class take three more steps, and halt (count "one . . . four," and "five" if on tip-toe).

For other forms of marching, see the "Standard Infantry

Tactics." A great many of the above commands and movements differ from those given in the book just named, but are more modern and hence worthy of adoption.

Whereas tip-toe marching is to be classified as a balance-movement, ordinary marching has more the characteristics of slow leg-movements. Running should always be followed by ordinary marching. Marching and running are usually put in somewhere in the middle of the table of exercises, and they should form part of every lesson in gymnastics. That they are necessary, no one will deny; for, as the Swedish proverb says. "As we stand and walk, so shall we act."

Close St. Pos. — Command, "*Feet — clóse!*" The balls of the feet are lifted from the ground, and, rotating on the heels, the feet are brought quickly together so that their insides touch, when they are again placed wholly on the floor. To resume fundamental position, command, "*Feet — ópen!*" The feet are turned out on the same principles as above. During the whole movement, the body should still gravitate forward as in fundamental position, and be as immovable as possible. This movement is used to correct the base — whenever you see that some pupils are standing with the feet unevenly turned out. If the movement does not immediately produce the desired effect, it should be repeated twice or more. In such a case, command, "*Feet close and open — óne! Twó!*" or "*The same — one! Twó!*" The close st. pos. is used as commencing position for various exercises.

Wing St. Pos. (Fig. 15). — In many movements it is desirable to fix the arms so as to gain better muscular isolation. This is done either by resting the hands on the hips or locking them together behind the neck. The former position is called wing pos. (wing st., wing sitt., etc.), on account of the arms resembling the wings of a bird. The command is, "*Hips—firm!*" The hands are quickly lifted, and grasp the waist firmly just above the hips. The fingers are held together in front and the thumbs are behind; the palms of the hands rest fully on the hips, and the elbows are slightly drawn backward. In this position the trunk is better isolated from the abdomen and legs, so that movements of parts on either side of the waist are more easily confined to prescribed parts alone. Besides, the respiratory muscles get a firmer support, respiration thus becoming

FIG. 15.—WING ST. POS.   FIG. 16.—REST ST. POS.

freer. (If the thumbs are in front and the fingers behind, the humeri become rotated forward, thus compressing the chest and pushing the shoulders forward — all faults.) A common fault is to draw the elbows too much backward.

Rest St. Pos. (Fig. 16), so called on account of its restful feeling. Command, "*Neck—firm!*" The hands are quickly lifted, and grasp the neck below the occiput, with fingers locked together, elbows carried well backward, and head remaining erect. A common fault is that of bringing the elbows forward, thus pushing the head forward and compressing the chest.

To prevent this it is allowable to let the fingers touch (by the tips) instead of locking them. In this position, the upper arms being lifted, the chest is much more expanded than in st. or wing st. positions; and the axillary artery being extended, the blood flows more easily to the arm. In many movements, the rest st. pos. is more powerful than the wing st. one, the lever for the weight (the trunk + the raised arms) being longer. The position should not be used until (by shoulder-blade movements) the shoulders have been rendered flexible enough to prevent a too faulty posture.

St. H. Backw. Flex. (Fig. 17). — Command, *"Head backward — bend!"* The head is bent slowly backward without changing the posture of the rest of the body. At the command, *"Upward — stretch!"* the head is slowly raised, the chin being first drawn in. When done for the first time, it is safest to make the pupils take wing pos., before doing the movement. Whenever the class have "hanging" heads, you order this movement two or three times in order to correct the posture. This is especially necessary after exercises which are likely to push the head forward.

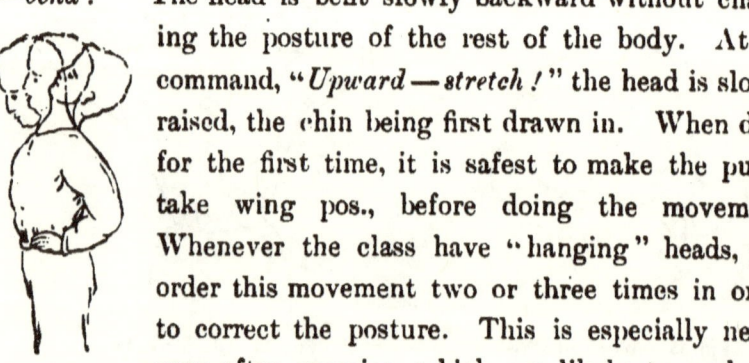

FIG. 17.—H. FLEX. BACKW. AND FORW.

St. H. Forw. Flex. (Fig. 17). — After some practice the head may also be bent forward at the command, *"Head forward — bend!"* *"Upward — stretch!"* The head is tipped forward, with the chin drawn in, as far as good posture will allow. This movement should be immediately followed by head flexion backward.

St. H. Sidew. Flex. (Fig. 18). — Command, *"Head to the left (r.) — bend!"* *"Upward — stretch!"* The movement, practised alternately to each side, should be ended by a head flexion backward.

St. H. Rot. (Fig. 19). — The head is rotated to either side at the command, "*Head to the left (r.) — turn!*" "*Head forward — turn!*" The head must not be turned so far that the shoulders are out of line. When done for the first time, the movement is executed slowly; later on, however, it takes place with utmost speed at the command, "*Head rotation, to the left (r.), right (l.), and forward — óne! Twó! Thrée!*" ... In this form it is used a great deal to keep the pupils busy in a posi-

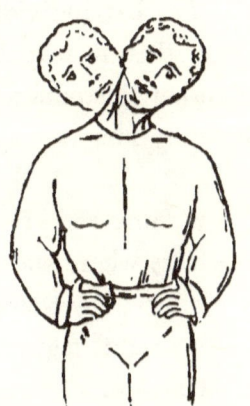
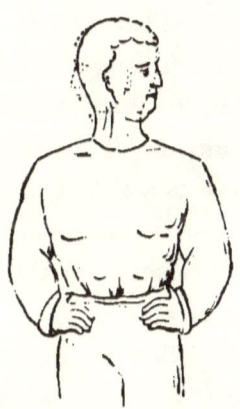

FIG. 18. — H. FLEX. SIDEW.   FIG. 19. — WG. ST. H. ROT.

tion which we wish them to hold for some time, so as to get the utmost effect out of it. When done slowly, the movement is suitably followed by head flexion backward, as the head involuntarily glides slightly forward when being turned to the side.

St. H. Rot. w. Flex. — Command, "*Head to the left (r.) — turn! (In this position) Head forward — bend! Head backward — bend! Upward — stretch!*" ... The head is bent obliquely in direction of the face and occiput.

All these movements tend to improve the posture of the head by strengthening the muscles of the neck.

LEG-MOVEMENTS.

In this class are included introductory leg-movements, as well as balance-movements and slow leg-movements. These are related as follows:

An exercise which has the distinct character of a balance-movement, — the difficulty of keeping the equilibrium making it a balance-movement, — and has been practised as such, ceases after some time to produce any effort towards maintenance of good balance, and then ceases for that individual (or class) to be a balance-movement. After that it may be used as an introductory exercise, beside other introductory leg-movements; or, if it be done in slow rhythm, it may sometimes serve as a slow leg-movement.

The slow leg-movements are either borrowed from the balance-movements, or they contain exercises where the passive extension of one group of muscles is produced by the activity of certain other groups, the active movement requiring but little effort. To go into an elaborate explanation of the effects of these movements, is not within the scope of this book. Suffice it to say, that these movements are used whenever the heart-beat has been too much accelerated by another movement, which they then immediately follow.

A progressive table of the balance-movements, and another of the slow leg-movements, will be found after the description of the leg-movements. The progressive arrangement of the introductory leg-movements is of less importance, and may be safely trusted to any teacher of average common sense.

In many movements, the nomenclature is simplified by merely naming the terminating position. When this is given,

MOVEMENTS OF THE SWEDISH SYSTEM 47

it naturally follows that the movement consists in taking this position, first to one side, then to the other.

Walk St. *a* **Pos.** (Fig. 20). — Command, "(*Hips — firm !*) *Left* (*r.*) *foot sideways forward — pláce !*" The foot is lifted and moved twice its own length in its own direction, and placed on the floor with a gentle pressure, the weight of the body being carried equally by both legs. "*Change feet — óne ! Twó !*" At "*one*" the foot is moved back into fundamental pos.; at "*two*" the other foot is moved in its own direction as above. Fundamental pos. is resumed at the command "*Posi-*

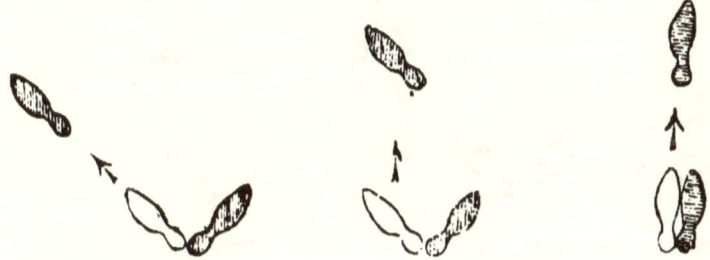

FIG. 20.—WALK ST. *a* POS.   FIG. 21.—WALK ST. *b* POS.   FIG. 22.—WALK ST. *c* POS.

tion !" The foot may be moved sideways backward on the same principles. (The position may then be named "Walk st. *a* pos. backw.")

Walk St. *b* **Pos.** (Fig. 21). — Command, "(*Hips — firm !*) *Left* (*r.*) *foot forward* (*backw.*) — *pláce !*" The foot is lifted and moved parallel with itself twice its own length forward, and placed on the ground with a gentle pressure, the body being carried equally by both legs. "*Change feet — óne ! Twó !*" on the same principles as above. The movement is done backward as well.

Walk St. *c* **Pos.** (Fig. 22). — Command, "(*Hips — firm !*) *Feet — clóse ! Left* (*r.*) *foot forward* (*backw.*) — *pláce !*"

The foot is moved twice its own length in its own direction. Care is taken to keep the feet pointing straight forward. " *Change feet — óne, twó !* " etc.

Stride St. Pos. (Fig. 23). — Command, "*Left (r.) foot sideways — pláce !* " The foot is moved parallel with itself, twice its own length straight to the side, and placed on the ground, so that the heels are still on the same line, parallel with a line through the shoulders; the feet are equally turned out, the knees straight, and the body carried equally by both legs. For beginners the movement is best given the following form: Command, "*Feet sideways place — óne ! Twó !* " At "*one*"

Fig. 23. — Stride St. Pos.   Fig. 24. — Crosswise St. Pos.

the left foot is moved its own length to the left; at "*two*" the right foot is moved its own length to the right. The fault of carrying the weight principally on one leg, which easily arises when only one foot is moved, is by this means prevented. When the movement has been done this way, the pupils take fundamental position at the command, "*Position — óne ! Twó !*" the movement beginning by the left foot being moved its own length to the right. When the right foot is placed beside the left one, the arms take fundamental position (if they were in any other position before).

Crosswise St. Pos. (Fig. 24). — Command, "*Left (r.) foot crosswise forward (backw.) — pláce !* " The foot is moved parallel with itself in direction of the other foot, and placed on the floor, the distance between the heels being a little less than twice the length of the foot. Care should be taken to keep the shoulders facing as before the foot moved. The body is carried equally by both legs. "*Change feet — óne! Twó!*" etc.

After some practice, each of the above movements can be practised in series of four counts. Command, for instance, "*Foot-placing sideways forward in four counts, beginning with the left* (r.) *foot — óne! . . . Fóur! The same, beginning with the right* (l.) *foot — óne! . . . Fóur!*" Practised in this way, these introductions improve the conception of time as well as that of space. Later on these positions are used as commencing positions for various exercises.

Fallout St. a Pos. (Fig. 25).— Command, "*To the left* (r.) *— fall óut!*" The body is allowed to fall in direction of the left (r.) foot, which is lifted from the floor, and moved three times its own length in its own direction; the left (r.) knee is bent to right angles, and remains so when the foot is placed on the floor. The forward foot thus carries most of the weight of the body. The trunk and backward leg are in a straight line, the shoulders facing as before, and both feet fully resting on the floor. "*Change feet — óne! Twó!*" (This needs no

FIG. 25. — FALLOUT ST. *a* POS.

description.) If the movement is to be done backward, the command is (for the position created above), "*Backward to the right* (*l.*) —*fall óut!*" In certain exercises this position is used with the trunk erect instead of inclining. To denote the difference, we then use the command, "*Left* (*r.*) *foot, large step, sideways forward* (*backw.*) — *pláce!*" In either case, when the movement is practised for the first time, arrange the class so that they stand one foot parallel to the boards of the floor, and the other at right angles to them (facing obliquely to the flank), as this will help the pupils to find the directions in which the feet should move.

Fallout St. *b* Pos. — This position resembles the previous one, except that the foot is moved parallel with itself forward (backward) (compare walk st. *a* as related to walk st. *b* pos.). Command, "*Left* (*r.*) *foot forward* (*backw.*) —*fall óut!*" or, if the trunk is to be erect, "*Left* (*r.*) *foot large step forward* (*backw.*) — *pláce!*" The easiest way to get this position correct is by moving the foot backward, especially if the trunk is to remain erect. If the backward foot is not fully resting on the floor, but has the heel raised, the position is called

Toe-support Fallout St. *b* Pos. (Fig. 149, p. 134). — The command for this is, "*Left* (*r.*) *foot, toe-supporting, large step backward — pláce!*" The distance between the heels is somewhat more than three times the length of the foot, and the forward knee is bent a little more than in the fallout st. *b* pos. If the backward foot is inserted between two of the stall-bars, the position is called

Foot Gr. Fallout St. *b* Pos. (Fig. 26). — To get into this position, the pupil stands one step away from the bars, turning his back to them. At the command, "*Left* (*r.*) *foot backward — pláce!*" one foot is inserted between two of the bars; and

MOVEMENTS OF THE SWEDISH SYSTEM   51

at the command, "*Forward — fall óut!*" the pupil jumps forward into position, the backward knee being stretched to its fullest extent. If the backward foot is lifted free from the floor, the position is called

Hor. Half St. Pos. (Fig. 28). — For this position the command is, "*Horizontal half-standing position, left (r.) leg backward — lift!*" The trunk and backward leg form a graceful curve, and the knee of the supporting leg is bent to nearly right angles.

FIG. 26. — FOOT. GR. FALLOUT ST. *b* POS.

Fallout St. *c* **Pos.** (Fig. 29) is like the fallout st. *b* pos., but the feet are pointed straight forward as in walk st. *c* pos. The distance between the heels should be three times the length of the foot, but may be less; if otherwise, the backward heel will be raised

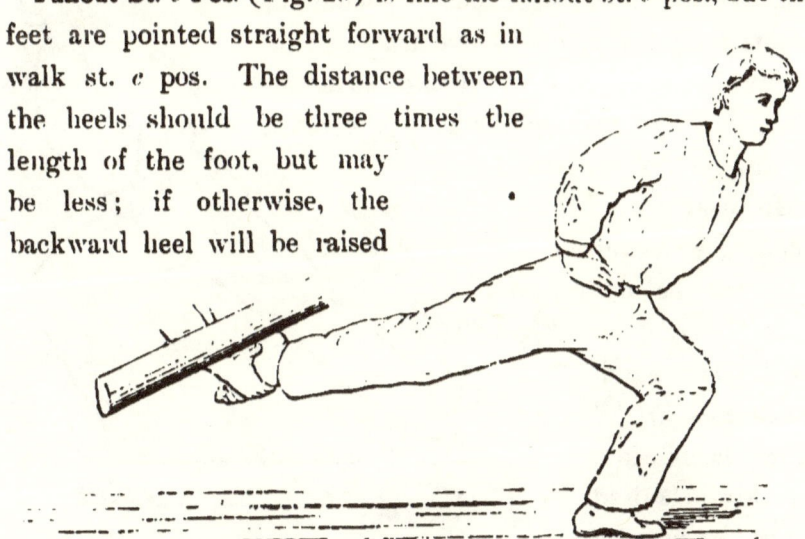

FIG. 27. — FOOT GR. FALLOUT ST. *b* POS.

from the floor. The fallout st. positions, except foot gr. fallout st. *b*, are not suitable for children (under twelve or fourteen years of age), on account of their inability to get the posture correct.

FIG. 28. — HOR. HALF ST. POS.

All the fallout st. positions belong among balance-movements, although the fallout st. *c* pos. has also the character of a slow leg-movement on account of the strong extension of the backward calf.

All the positions from walk st. *a* to fallout st. *c* inclusive, are to be practised with "hips — firm," which command should immediately precede these movements. Thus each of these positions should have the prefix "wing," making them: wg. walk st. *a*, wg. hor. half st., etc.

In nearly all of the above positions, the movement of heel-elevation can be performed.

St. Heel-elev. (Fig. 30). — Command, "*Heels — lift!*" The heels are quickly raised from the ground, and the pupil

FIG. 29. — FALLOUT ST. *c* POS.

stands as high on tiptoe as possible. This position is called **toe standing**. "*Heels — sink!*" The heels are again lowered, with moderate speed, without tipping the weight of the body backward. On the same principles, the following movements are executed: —

Close st. heel-elev. (Fig. 31).
Stride st. heel-elev. (Fig. 32).
Walk st. *a* heel-elev.
Walk st. *b* heel-elev. (Fig. 33).
Walk st. *c* heel-elcv.
Crosswise st. heel-elev.

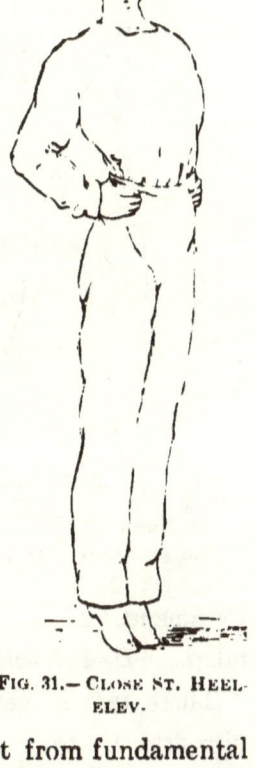

FIG. 31.— CLOSE ST. HEEL-ELEV.

In the last five positions, the movements can be arranged in series of four counts. For instance: "*Foot-placing sideways forward with heel-elevation, left (r.) foot — óne! ... Fóur!*" "*The same, right (l.) foot — óne! ... Fóur!*" etc. At the first count, the foot is moved out from fundamental position; the heels are raised at the second and lowered at the third; and at the fourth the foot resumes fundamental position.

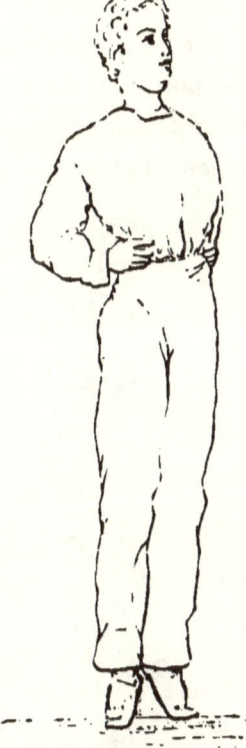

FIG. 30.— CLOSE ST. HEEL-ELEV.

To increase the difficulty of balance, the movements, which are first done in wing. pos., are later done with the arms in yard, rest, and stretch positions (see heaving-movements), in the order mentioned. (By raising the arms, the centre of gravity is lifted higher above the base, making the equilibrium less stable.) These movements are soon used as introductions.

Fallout St. *a* **Heel-elev.** (Fig. 34). — Command, "(*Hips — firm!*) *To the left* (*r.*) — *fall óut! Heel-elevation — óne! Twó!*" Here only the forward heel is raised — not both, as described by some authors. By the raising of the heel, the forward knee is bent still more. In a similar manner is executed **Fallout St.** *c* **Heel-elev.**—

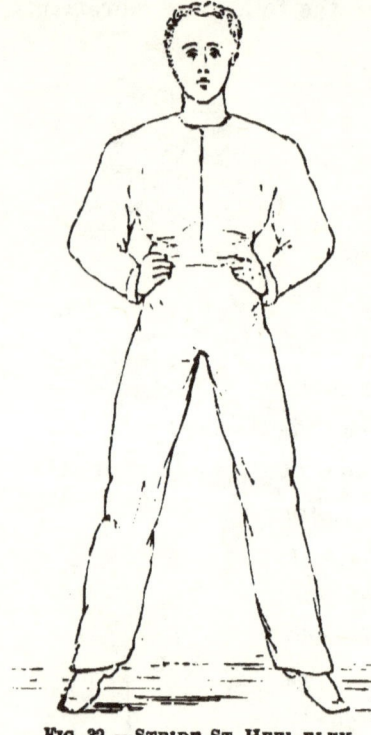

FIG. 32.— STRIDE ST. HEEL-ELEV.

FIG. 33. — WG. WALK ST. *b* HEEL-ELEV.

Command, after fallout st. *c* position is taken, "*Heel-elevation — óne! Twó!*"

These two movements, always done with hips firm, are to be considered chiefly as slow leg-movements: in the first, the quadriceps extensor of the forward leg is forcibly active (eccentric contraction); in the second, the gastrocnemius of the backward leg (passive extension). Heel-elevation in fallout st. *b* position is possible, but has but little value, the effort producing a hardly noticeable change in the muscular activity already present in the commencing position.

Alternate Toe-elev. — Command, *"Alternate toe-elevation — óne! Twó!"* . . . The heels remain on the ground, and the balls of the feet are alternately lifted and again put down with a slight pressure, one foot moving up as the other one moves down. (The speed is about seventy movements a minute.) This is done with or without hips firm. After some practice, the command is changed to *"Alternate toe elevation — stárt! . . . Stóp!"* the left foot being the one first lifted. The shoulders should not be tipped from side to side, as is apt to be the case; nor should the body cease to gravitate forward. When done as last described, the movement is very efficient in giving the pupils a true conception of rhythm. To increase this effect, they may be allowed to put the foot down a little more forcibly than is usually the case. This is also a good means of counteracting the tendency of the pupils to increase the speed of motion. Children usually find it very amusing to do this movement; and in their gymnastics it is used a great deal as an introduction.

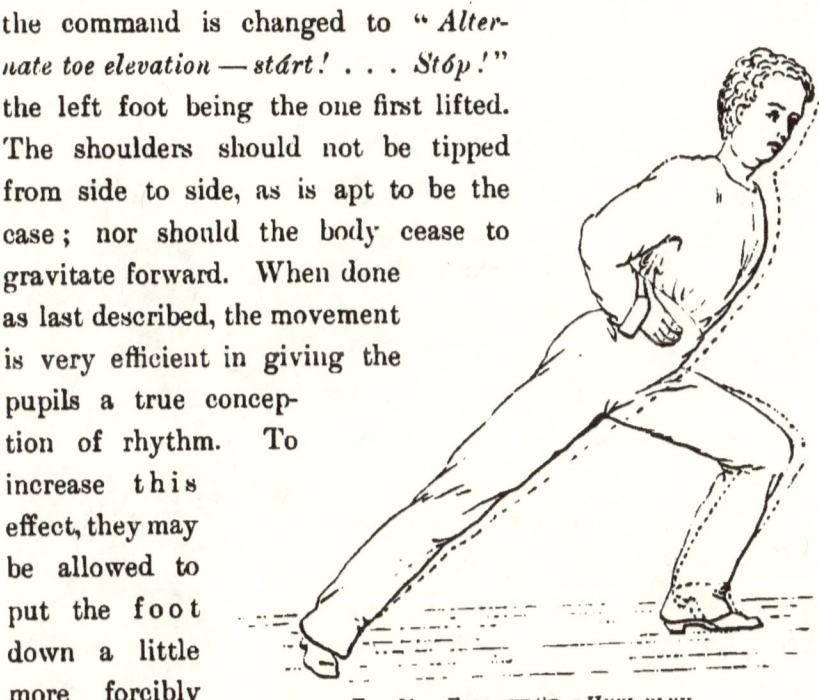

Fig. 34. — Fallout St. *a* Heel-elev.

Alternate Heel and Toe-elev. — Command, *"Alternate heel and toe elevation — stárt!"* (or *óne! Twó! . . .*). 1. The heels are raised and lowered. 2. The balls of the feet are raised and lowered, and the movement is repeated until the command, "*Stóp!*" is given. The body should not be swaying backward and forward. For beginners the movement may be done in the reach grasp st. pos. (Fig. 35), in which this swaying is not possible.

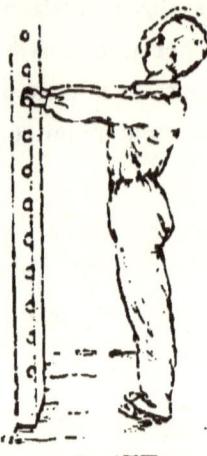

FIG. 35. — REACH GR. ST. ALT. HEEL AND TOE ELEV.

These two movements are slow leg-movements, the passive extension of the calf (alternating with its active contraction) propelling the blood into the legs, *i. e.*, from the heart.

In some of the above positions both knees can be bent, which movement we call double knee-flexion (= 2 Kn. flex.).

St. 2 Kn. Flex. (Fig. 36). Command, "*Heels — lift! Knees — bend! Upward — stretch! Heels — sink!*" The knees are bent to right angles; when so bent, the position is called **courtesy standing**. A common fault in this position is to tip the trunk forward and the seat backward. The knees should be pointing in the directions of the feet. This movement is a true balance-movement, to be done with moderate speed; but, as it contains the intermediate positions of jumping, after a while the speed and rhythm should vary so that

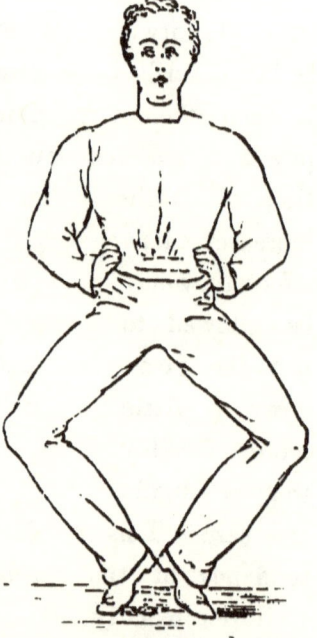

FIG. 36. — ST. 2 KN. FLEX. COURTESY ST. POS.

the pupils may become thoroughly balanced in each one of these postures. At such times, the command is, "*Prepare to jump* (or *preparation for jumping*)—*óne! Twó! Thrée! Fóur!*" a pause being made between any two of the executory commands. In this form — in spite of the speed or motion — the movement has also the character of a slow leg-movement, although when

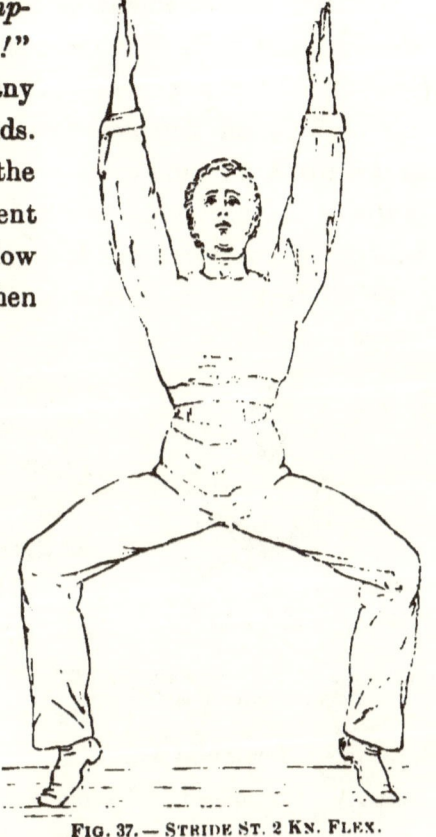

FIG. 37. — STRIDE ST. 2 KN. FLEX. COURTESY STRIDE ST. POS.

used as such it is usually done in very slow rhythm.

On the same principles command and execute

Stride st. 2 Kn. flex. (Fig. 37).
Walk st. *b* **2 Kn. flex.**
Crosswise st. 2 Kn. flex.

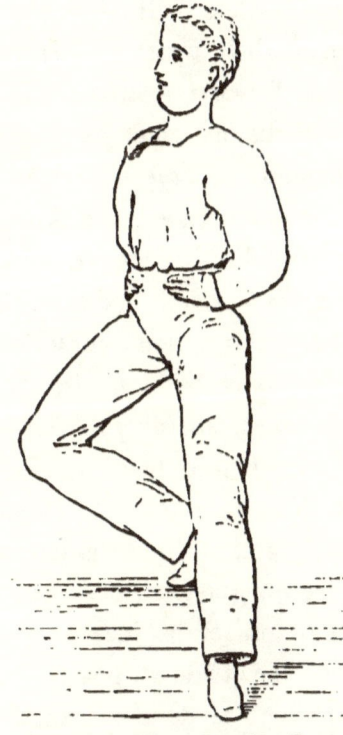

FIG. 38. — WALK ST. *a* KN. FLEX.

Walk St. *a* **Kn. Flex.** (Fig. 38). The command is the same as for double knee flexions, but the movement differs from these in that the backward knee is bent the most, the weight of the body being carried chiefly by the backward leg. If both knees were equally bent, the difference between this movement and walk st. *b* 2 Kn. flex. would not be great enough to warrant the use of different commencing positions. Besides, when done as described, the intermediate position (as shown in the figure) is almost identical with the fundamental position in fencing, for which this movement will thus prepare the way.[1]

St. 2 Kn. Flex. to Sitt. Pos. (Fig. 39). Command, "*Heels — lift! Knees — bend! Sit!*" The knees are bent as far as muscular resistance will allow; the heels come slightly apart. This position is called **courtesy sitting**. The trunk should be as erect as usual. At the command, "*Knees — stretch!*" or "*Upward — stretch!*" the knees are again stretched to toe st. pos.; and at "*Heels — sink!*" the heels are lowered. After some practice,

FIG. 39. — 2 KN. FLEX. TO SITT. POS. COURTESY SITT. POS.

especially for children, the movement may be done in five counts. Command, "*Double knee flexion to sitting position — óne! . . . Five!*" (1. Heels lift. 2. Knees bend to 90°. 3. Knees bend to sitting position. 4. Knees stretch. 5. Heels sink.) "*The same — óne! . . . Five!*" etc. When the movement is done for the first time, the reach grasp st. pos. (compare Fig. 35) may be used to prevent loss of balance.

The movement can also be done from the stride st. pos. : —

Stride St. 2 Kn. Flex. to Sitt. Pos., which is done on exactly the same principles.

[1] Fencing may be considered as gymnastics for advanced classes; hence, one should lead to the other.

In all these 2 Kn. flexs. (from st. 2 Kn. flex. to the one just named inclusive), a progression is made by having the arms in wing, yard, stretch, or rest positions; by staying in the courtesy st. or sitt. position, while a head rotation or 2 A. ext. (for which see heaving-movements) is being executed; or by combining the knee flexion with an arm extension. To enumerate all the various movements to which these combinations give rise, would take too much space. The following movement, as being the most complicated one, may however be described:

Bend Toe St. 2 A. Ext. and 2 Kn. Flex. Command, "*Arms upward bend!* (see heaving-movements) *and heels — lift! Arm extension upward with knee flexion — óne! Twó!*" ... 1. The arms are stretched up, and the knees bent. 2. The arms are bent, and the knees stretched. This is especially suitable for advanced pupils.

In the various walk standing positions, as well as in stride st. pos., the double knee flexion can be done in series of six counts. Thus: "*Foot-placing sideways with double knee flexion, left (r.) foot — óne! ... Síx!*" 1. Left foot sideways place. 2. Heels lift. 3. Knees bend. 4. Knees stretch. 5. Heels sink. 6. Position. "*The same, right (l.) foot — óne! ... Síx!*" etc. To be done only by advanced classes. After a while the teacher leaves off counting, when the executory command is, "*Stárt!*" the pupils going through the whole movement, and taking the rhythm from those who stand foremost.

Toe Sup. Walk St. *b* Kn. Flex. (Fig. 41). Command, "*Left (r.) foot, toe supporting, backward — pláce!*" The position is like walk st. *b* (Fig. 40), except that the backward heel does not touch the ground. "*Right (l.) knee — bend!*" The forward knee is slowly bent until the heel of the (straight) backward leg touches the floor. This causes passive extension of the backward calf, making the exercise a slow leg-movement. It is apt to produce a very faulty posture (on account of the difficulty of attaining muscular isolation, or confining the movement to prescribed parts alone), and consequently must not be given to children.

Instep Gr. St. Kn. Flex. (Fig. 43). The pupil stands at a distance of twice the length of his foot from the stall-bars or other apparatus (Fig. 42). Command, "*Left (r.) foot backward — pláce!*" The foot is inserted between the third and fourth bars, the knee of this leg being bent. "*Right (l.) knee — bend!*" The knee of the carrying leg is bent as far as the extensibility of the quadriceps of the other leg will allow, the trunk remaining erect with the hips brought well forward. Besides being a slow leg-movement, this exercise cultivates the flexibility of the knee, and thus improves the elasticity in jumping.

These two movements can also be done from rest or str. st. positions.

Heel Gr. Half St. Kn. Flex. The pupil stands facing the stall-bars; and, at the command "*Left (r.) foot forward — pláce!*" he puts this foot between two bars at hip-height (or lower), or he rests his heel on some other convenient apparatus, as shown in Fig. 44. At the command "*Right (l.) knee — bend!*" the knee of the carrying leg is bent, while the supported one remains straight, the trunk being erect or slightly inclined forward. The exercise produces a passive extension of the whole under side of the straight leg. A similar effect is received if both knees remain straight, while the trunk is bent forward, which exercise is named

MOVEMENTS OF THE SWEDISH SYSTEM 61

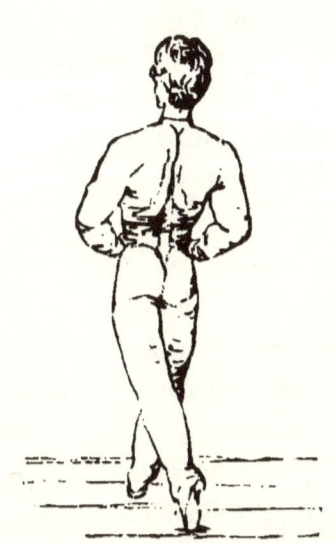

FIG. 40.— TOE SUP. WALK ST. b POS.

FIG. 41.— TOE SUP. WALK ST. b KN. FLEX.

FIG. 42.— INSTEP GR. ST. POS.

FIG. 43.— INSTEP GR. ST. KN. FLEX.

Heel Gr. Half St. T. Forw. Flex. (Fig. 44).—The commencing position is the same as above. The command for the movement is, "*Trunk forward — bend! Upward — stretch!*" (Compare arch-flexions.) A similar effect is also received by

FIG. 44.—HEEL GR. HALF. ST. T. FORW. FLEX.

Str. Walk St. *b*, T. Forw. Downw. Flex. (Fig. 45).—Command, "*Left (r.) foot forward and arms upward — stretch!*" The forward foot is braced against the horizontal bar, resting on the floor (or against a platform, etc.). "*Forward, downward — bend!*" The trunk is bent as far as good posture will allow. Both legs remain straight, and the backward heel must not leave the floor. "*Upward — stretch!*"

Half Wing, Half Yard Gr. Half St. Kn. Flex. (Fig. 46).—Command, "*Right (l.) side to the bars. Left (r.) hand hip firm, right (l.) hand grasp at hip-height! Left (r.) leg forward — lift!*" The leg is lifted straight forward to nearly horizontal position. "*Heel — lift! Knee — bend! Knee — stretch!*" ... The knee of the carrying leg is bent to right angles (or to sitting position), while the lifted leg remains

FIG. 45.—STR. WALK ST. *b*, T. FORW. FLEX.

straight and horizontal, and the trunk erect. A common fault is to tip the body backward. This movement can also be done without heel-elevation. Instead of the stall-bars, the back of a chair, the shoulder of another pupil, etc., may serve for support.

Reach Gr. Half St. Kn. Flex. (Fig. 47). — The pupils grasp the horizontal bar, which they are facing (or the hands of the other pupils, as shown in the figure). Command, "*Left (r.) leg forward — lift! Heel — lift! Knee — bend!*" . . .

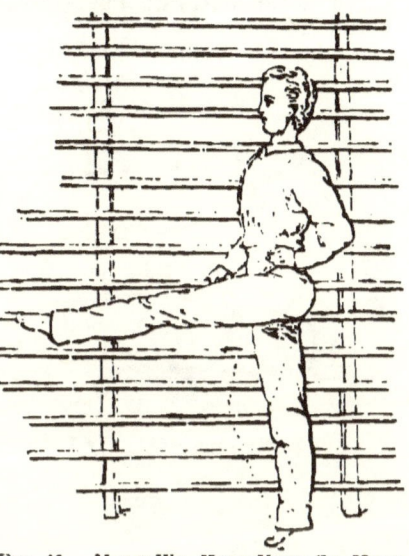

FIG. 46. — HALF WG. HALF YARD GR. HALF ST. KN. FLEX.

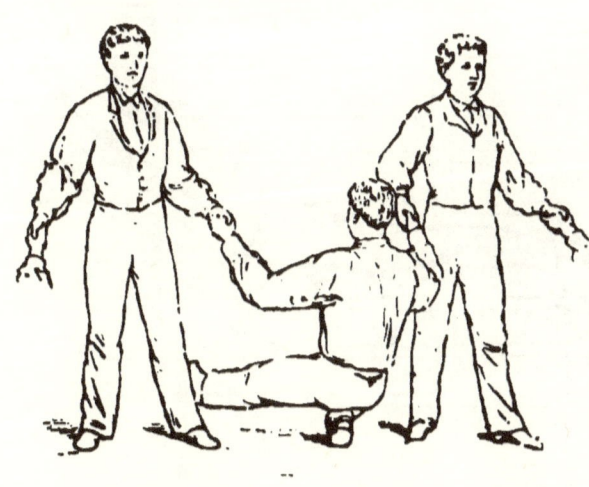

FIG. 47. — REACH GR. HALF ST. KN. FLEX.

etc. The movement is executed on the same principles as the one just described, — with or without previous heel-elevation, and to 90° or to sitting position. Without support for the hands (entirely free-standing), the movement is performed in a like manner. It is then called

Reach Half St. Kn. Flex. (Fig. 48). — This movement is always done without heel-elevation. Command, "*Arms and left (r.) leg forward — lift! Knee — bend!*" etc.

Half Yard Gr. Hor. Half St. Kn. Flex. (Fig. 49). — One hand grasps the stall-bars, the other takes hip firm. Command, "*Horizontal half standing position, right (l.) leg backward — lift!*" (Performed as described above.) *Left (r.) knee — bend!*" The knee is bent as far as possible. "*Knee — stretch!*" The knee resumes its former flexion (about 90°). This can also be done with previous heel-elevation and with one arm extended upward (hlf. str. yard gr. hor. half st. pos.) as shown in the figure.

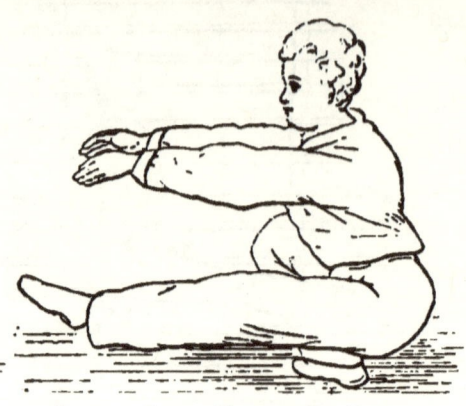

FIG. 48. — REACH HALF ST. KN. FLEX.

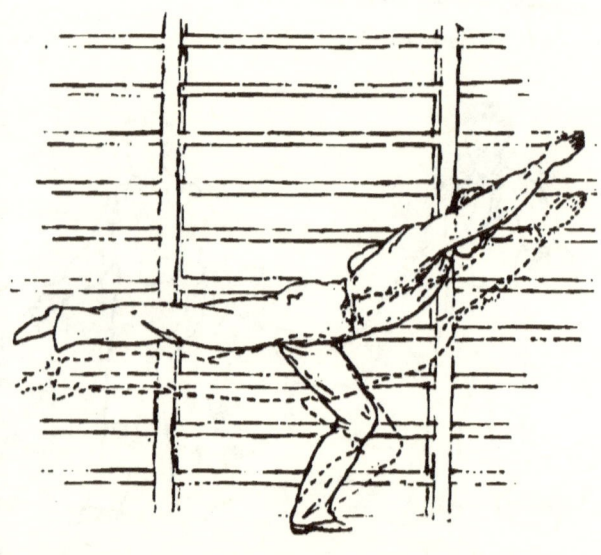

FIG. 49. — HALF YARD GR. HOR. HALF. ST. KN. FLEX.

The movement can also be free-standing, and is then called **Horizontal Half St. Kn. Flex.** — It is executed as above, always without heel-elevation. It may be done from wing, half str. or str. pos. When only one arm is extended upward, the position may be either left (r.) arm up, and left (r.) leg lifted backward, or *vice versa*. The last five movements are both balance-movements and slow leg-movements. Improving the flexibility of the knee, they also improve the elasticity required in jumping. The str. and half str. hor. half st. Kn. flex. are also shoulder-blade movements, on account of the difficulty of keeping the arms in correct position.

Crook Half St. Pos. (Fig. 50).—Command, "*Left (r.) knee upward — bend!*" The leg is slowly lifted forward and upward until the hip and knee are bent at right angles; the instep is stretched, so that the toes are pointing downward; the knee is pointing sideways forward. Common faults are to tip the shoulders to the opposite side, to incline the trunk backward, and to bend the knee of the leg that carries the body. "*Change feet — one!*

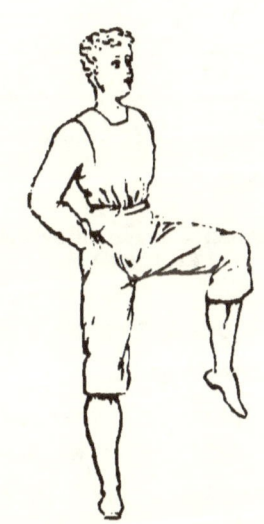

FIG. 50. — WG. CR. HALF ST. POS.

Two!" 1. The lifted foot is put down. 2. The other leg is raised as above. For beginners the change of feet is done quickly and rhythmically, as this is easier to do (seventy to a hundred movements a minute). When done this way, the knee is lifted somewhat higher. In this form, the exercise forms a substitute for marching, when space does not allow the latter; and as marching, it can also be done on tip-toe — when a slight motion forward is allowed — and in double-quick time.

Beginners, who are not able to take the crook half st. pos.

correctly, may for a while be allowed to use support, the exercise then being

Reach Gr. St. Alternate Kn. Flex. Upw. (Fig. 51), which is done on exactly similar principles.

In the crook half st. pos., various exercises are done in progression, as enumerated below: —

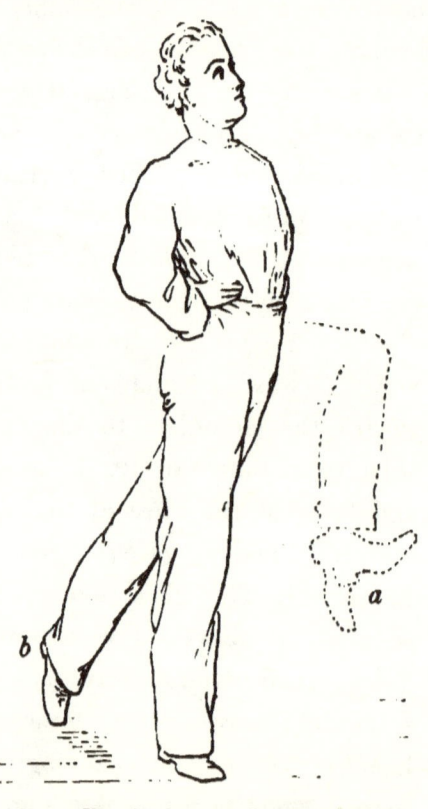

Fig. 51.— Reach Gr. St. Alt. Kn. Flex. Upw.

Crook Half St. F. Flex. and Ext. (Fig. 52 *a*).— Command, "*Left (r.) knee upward — bend! Foot flexion and extension — óne! Twó!*" ... The figure is sufficient explanation.

Crook Half St. Kn. Ext. Backw. (Fig. 52 *b*). — Command, "*Left (r.) knee upward — bend! Knee backward — stretch! Upward — bend!*" ... The knee is extended as far backward as good posture will allow, — *i.e.*, without bending the trunk forward.

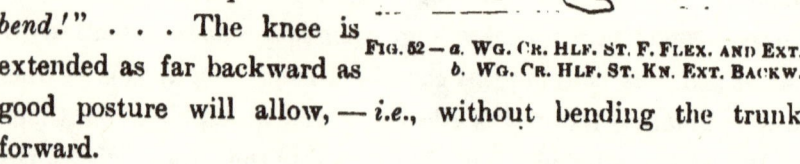

Fig. 52 — *a*. Wg. Cr. Hlf. St. F. Flex. and Ext. *b*. Wg. Cr. Hlf. St. Kn. Ext. Backw.

Crook Half St. Kn. Ext. Forw. (Fig. 53). — Command, . . . "*Knee forward — stretch! Bend! Stretch!*" . . . Common faults are, to lean over backward, and to bend the knee of the carrying leg. To correct these, supports (Fig. 54) may be allowed for a short time.

Crook Half St. Kn. Abd. — Command, . . . "*Knee abduction — one! Two!*" . . . 1. The raised leg is moved as far sideways as good posture will allow. 2. It is moved back to original position. The shoulders (and trunk) should not be twisted from side to side.

FIG. 53. — CROOK HALF ST. KN. EXT. FORW.

In all these balance-movements, progression is made from wing st. to yard st., rest st. and str. st. positions, as enumerated; yet so that in Kn. ext. forw., the str. st. pos. precedes the rest st. pos., as the former one is less apt to produce a faulty posture in this movement.

FIG. 54. — REACH GR. HALF ST. KN. EXT. FORW.

St. L. Elev. (Fig. 55). Command, "*Left (r.) leg sideways — lift!*" The weight of the body is removed to the right (l.) leg, and the straight left (r.) leg is lifted sideways as far as possible. The shoulders are kept level, and the trunk erect. "*Change feet — one! Two!*" . . . 1. The lifted foot is again put down beside the other one. 2. The other leg is raised as above. The movement is made more difficult by using either of the wing, rest, str., or yard st. positions in the order enumerated. The yard pos. is put last, because, in this exercise, it is more difficult to maintain than any of the others. Finally the movement can be done from toe st. pos., which necessitates that the arms are in stretch pos. Command, "*Arms upward — stretch! and heels — lift! Leg elevation — one! Two!*" . . . The movement is executed with more speed in this than in the other positions.

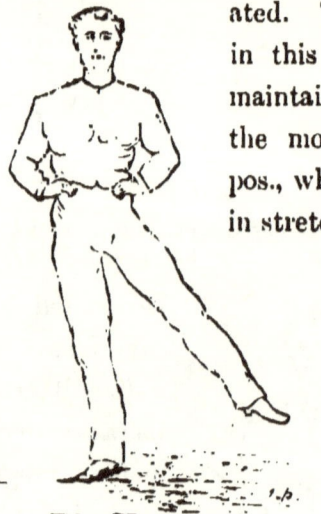

Fig. 55. — Wg. St. L. Elev.

Leg-elevation sideways is also a lateral trunk-movement, since it brings into play the muscles around the waist; but, when used as such, it is best done from other commencing positions.

Among leg-movements, we also classify marching and running. Marching in ordinary time is one of the best slow leg-movements. Marching on tip-toe is an excellent balance-movement, and is one of the best exercises to correct a faulty posture. Command, "*Hips firm and feet — close! Heels — lift! Slowly forward march — one! Two!*" . . . This movement is also done with neck firm or arms upward stretch, and later on with the feet turned out. The shoulders should be kept level throughout the exercise, which is then one of the best means of correcting the bad habit of tipping the shoulders from side to side when walking. Taking place on tip-toe, the movement also counteracts the tendency we all have to walk on our heels: it teaches us to land on the balls of our feet, making the steps springy and graceful.

As balance-movements, we can use various forms of balancing on the horizontal bar (Fig. 56), an exercise which can be done backward as well as forward, and is too familiar to us all to need any description. The bar (at first at knee-height or lower) should have the flat side up until the pupils have gained some proficiency in the movement. After that the round side may be turned up,

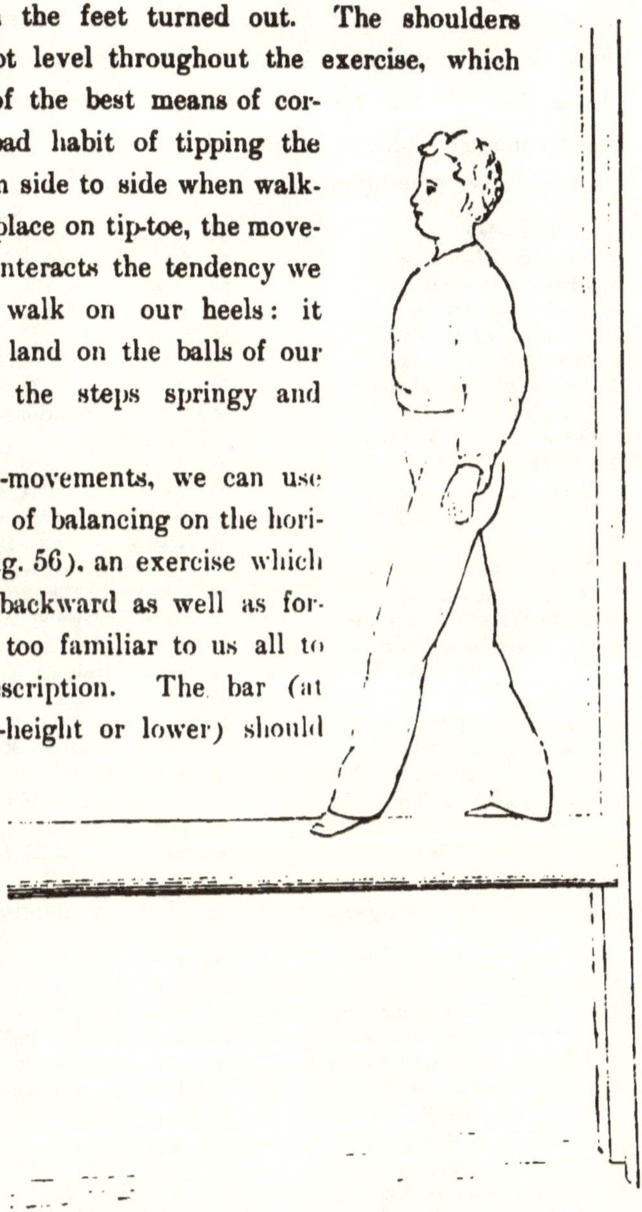

FIG. 56. — BALANCE WALKING ON HORIZONTAL BAR.

and the bar may be gradually raised. In the progressive

arrangements of exercises which follow the descriptions of each class, *all* the movements have not been put in; but the teacher can easily fit in the few remaining ones for himself, when he has the majority already arranged.

Progressive arrangement of balance-movements: —

1. St. Heel-elev.
2. Stride St. Heel-elev.
3. Close St. Heel-elev.
4. Walk St. *a* Heel-elev.
5. Walk St. *b* Heel-elev.
6. Walk St. *c* Heel-elev.
7. Crosswise St. Heel-elev.
8. Wing St. 2 Kn. Flex.
9. Wing Walk St. *b* 2 Kn. Flex.
10. [Reach Gr. St. Alt. Kn. Flex. Upw.]
11. Wing St. Alt. Kn. Flex. Upw.
12. Wing Crosswise St. 2 Kn. Flex.
13. Fallout St. *a* Pos.
14. Wing Courtesy Sitt. Pos.
15. Wing Stride St. 2 Kn. Flex.
16. Wing Walk St. *a* Kn. Flex.
17. Str. Walk St. *a* Kn. Flex.
18. Wing Cr. Half St. F. Flex. and Ext.
19. Fallout St. *b* Pos.
20. Toe St. H. Rot. (children).
21. Wing Cr. Half St. Kn. Ext. Backw.
22. Yard Toe St. 2 Kn. Flex. to Sitt. Pos.
23. Str. St. 2 Kn. Flex.
24. Wing St. L. Elev.
25. Str. (Rest) Stride St. 2 Kn. Flex.
26. Wing St. Kn. Ext. Forw.
27. Wing Cr. Half St. Kn. Abd.
28. Bend Toe St. 2 A. Ext. & 2 Kn. Flex.
29. Wing Toe St. Alt. Kn. Flex. Upw.
30. Wing (Yard) Courtesy St. H. Rot. (children).
31. Wing Courtesy Sitt. H. Rot.
32. Courtesy Sitt. 2 A. Ext.
33. Rest St. Kn. Flex. Upw.
34. Rest St. Kn. Ext. Backw.
35. Str. Stride Courtesy Sitt. 2 A. Ext.
36. Rest St. L. Elev.
37. Str. Crosswise St. 2 Kn. Flex.
38. Str. Cr. Half St. Kn. Ext. Backw.
39. Str. Cr. Half St. Kn. Ext. Forw.
40. Str. St. L. Elev.
41. Str. (Rest) Cr. Half St. Kn. Abd.
42. Rest Cr. Half St. Kn. Ext. Forw.
43. Hor. Half St. Pos.
44. Reach Half St. Kn. Flex.
45. Yard St. L. Elev.
46. Str. Toe St. L. Elev.

Progressive arrangement of slow leg-movements: —

1. St. Alt. Toe-elev.
2. Fallout St. *c* Pos.
3. Reach Gr. St. Toe and Heel Elev.
4. Str. Walk St. *c* Heel-elev.
5. Toe Sup. Walk St. *b* Kn. Flex.
6. Slow 2 Kn. Flex. to Sitt. Pos.
7. Rest Walk St. *a* Kn. Flex.
8. Instep Gr. St. Kn. Flex.
9. Fallout St *a* Heel-elev.
10. Heel Gr. Half St. Kn Flex.
11. Half Yard Gr., Half St. Kn. Flex
12. Str Stride St. 2 Kn. Flex
13. Bend Courtesy Sitt. 2 A Ext.
14. Fallout St. *c* Kn. Flex.
15. Bend Toe St. 2 A. Ext. & Slow 2 Kn. Flex. to Sitt. Pos.
16. Str. Walk St. *b* T. Forw. Flex.
17. Yard Courtesy Sitt. 2 A. Elev.
18. Str. Walk St. *a* Kn. Flex.
19. Reach Gr. Half St. Heel-elev. and Kn. Flex.
20. Reach Half St. Kn. Flex.
21. Half Str., Half Yard Gr. Hor. Half St. Heel-elev. and Kn. Flex.
22. Str Hor. Half St. Kn. Flex.

ARCH-FLEXIONS.

By arch-flexions proper, we mean backward flexions of the trunk, with the arms in stretch position and the hands grasping a stall-bar or other firm support, as well as movements executed in this position (called str. gr. arch st. or gr. arch st. pos.). In this class must be included all the free-standing trunk flexions backward and forward, which are introductions to the corresponding movements with apparatus. In total absence of apparatus, we might also embrace here those trunk flexions backward and forward which are done after previous rotation of the trunk, and which, strictly speaking, belong among the lateral trunk movements (among which they are described in this book).

The effects of the arch-flexions are to draw the lower ribs apart, thus vaulting the chest forward: to improve the extensibility of the abdomen, and to straighten the dorsal region of the spine. They are nearly related to certain forms of the abdominal exercises, the essential difference being that in these the lower ends of the active muscles are fixed, whereas in the arch-flexions proper these are the moving ends, the upper ones being fixed.

In all the backward flexions described below, the whole spine should partake of the flexion — that is, the back should not be bent merely at the waist. For this reason, let the head begin the movement backward, and let it be the last in moving forward. Respiration should be free, and an occasional "*Breathe!*" from the teacher is not out of place. Another excellent bit of advice when backward flexion takes place is, "*Look at the opposite wall!*" (or, if at the stall-bars, "*Look at the bars!*"); this reminds the pupil to let the head also move backward. In forward flexion, you say, "*Look at the ceiling!*" which order prevents the head from losing its good posture. Whether the flexion is backward or forward, the knees should be kept straight.

St. T. Flex. Backw. and Forw. — Command, "*Trunk backward — bend!*" The trunk is bent slowly backward (Fig. 57). The position thus arising is called **arch st. pos.** "*Upward — stretch!*" The trunk resumes fund. pos. "*Trunk forward — bend!*" The trunk is bent slowly forward with fully expanded chest (Fig. 58). This position is called **stoop st. pos.** "*Upward — stretch!*" The trunk resumes fund. pos. The movement is also done in the close st., walk st. *b*, and stride st. positions. The last-named position may precede the others, as the muscular resistance in the legs is not so great in this as in the other positions, hence the flexion is deeper and easier. The arms may be either in wing, rest, or str. pos., except that trunk flexion forward with neck firm is used only for advanced pupils, as this movement is exceedingly difficult.

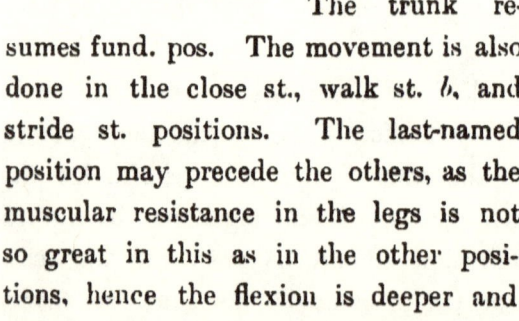

FIG. 57. — T. FLEX. BACKW.

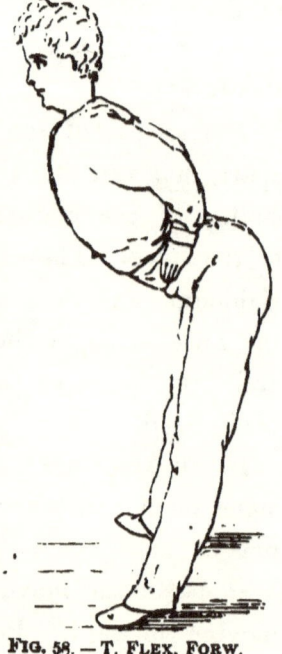

FIG. 58. — T. FLEX. FORW.

For children, the half str. st. position (Fig. 62,) may also be used. Thus we have

Wing	Stride st.	
Half str.	St.	T. flex., backw. and forw.
Rest	Walk st. *h*	
Str.	Close st.	

T. Flex. Forw. and Downw. (Fig. 59). — As soon as the flexion forward can be well executed, it may also be done downward. The arms should always be in str. pos., as otherwise the movement causes compression of the chest. Command, "*Arms upward — stretch!*" (for which, see heaving movements.) *Trunk backward — bend!* . . . *Forward, downward — bend!*" The trunk is bent as far as a good posture of the head, arms, shoulders, and chest will allow, the hands reaching out as far as possible. It is only after very long practice that the hands touch the ground. "*Upward — stretch!*" The trunk resumes fund. pos., which movement commences at the hands, which are the first to move upward backward. The small of the back should be concaved and the dorsal part flattened

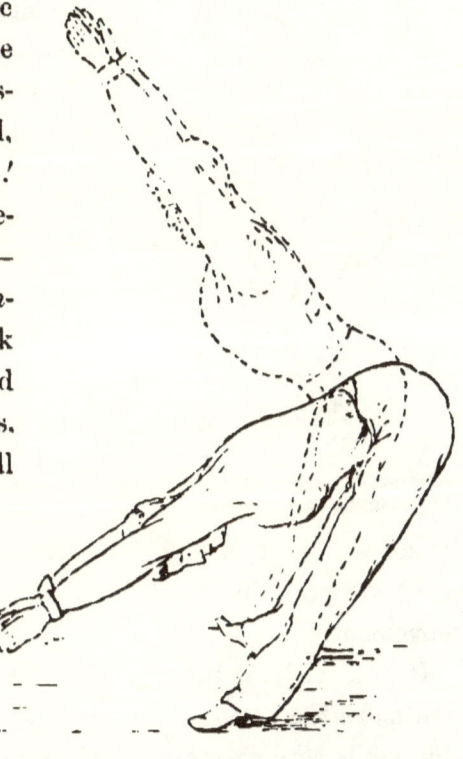

Fig. 59. — Str. Stride st. T. Flex. Forw. and Downw.

during the movement, so that the back is not convexly curved (in the familiar attitude of a frightened cat), but the chest must be thoroughly expanded throughout the movement. The exercise is also done in walk st. *b* or stride st. pos., the last-named one usually preceding the others. For children and beginners, the downward flexion may be done with less attention to good posture, and the hands made to touch the floor; for, the arms being extended upward (as related to the trunk), the compression of the chest is never excessive. Done in this way, the final position looks as in Fig. 60, which for an advanced pupil is decidedly faulty. When thus done, the movement is executed a little more quickly.

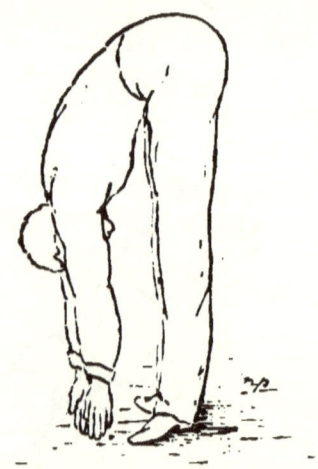

FIG. 60. — FAULTY FINAL POSITION IN TRUNK FLEXION DOWNWARD.

If, after downward flexion, you wish the pupils to stretch to stoop st. pos., you command, "*Forward — stretch!*" when the pupils take the position marked by dotted lines in Fig. 59. This mode of dividing the " upward stretch " in two parts is a good safeguard against faulty movement.

If you wish to increase the effect of the backward flexion, you make the pupils stay in arch st. pos. and do an arm extension while staying there. The movement is called

Str. Arch St. 2 A. Ext. Upw. — Command. "*Arms upward — strétch! Trunk backward — bend! Arm-extension upward — one! Twó!* . . . (for which movement, see heaving-movements.) *Trunk upward — stretch!*" etc. The movement may also be done from bend arch st. pos., which is a little easier. From this position each arm may be alternately extended, the final position then being half str. half bend arch st. pos. In this way it is, however, applied only to children. The movement can also be done from stride st. and walk st. *b* pos., the names then being

Str. Arch Stride St. 2 A. Ext., and

Str. Arch Walk St. *b* **2 A. Ext.** — In this position a still further progression can be made by combining the arm extension with a change of feet; thus:

Str. Arch Walk St. *b* **2 A. Ext. and Change of F.** — Command, "*Left (r.) foot forward place and arms upward stretch — one! Two!*" 1. Arms upward bend. 2. Arms upward stretch, and left (r.) foot forward place. "*Trunk backward — bend! Arm extension and change of feet — one! Two!*" ... 1. Arms upward bend, and place the forward foot beside the backward one. 2. Place the other foot forward, and stretch the arms upward. This exercise is quite powerful.

All backward flexion should be immediately followed by a forward (and downward) flexion.

Str. Gr. Arch St. Pos. (Fig. 61). — This is the first real arch-flexion. The class stands at the bar-stalls, one pupil at each, with his back turned to the bars. Command. "*A short step forward — márch! Arms upward — strétch! Trunk backward — bend!* (*Look at the bars!*)" The trunk is bent backward until the hands touch a bar; at first the hands merely rest on the bar, later on they grasp it with the palms under the bar, the thumbs in front of it, and both hands holding the same bar. If one hand grasps the bar below that held by the other, the position is called **oblique gr. arch st.** (Fig. 75, p. 84). If only one arm is up, the position is called **half str. gr. arch st.** (Fig. 62) (suitable for children).

In absence of apparatus, the wall may serve for support, the hands resting against it (Fig. 63). Or the pupils of the second rank may serve as supports (Fig. 64). The command then is, *"First rank, arms upward — stretch! Trunk backward — bend! Second rank, support — stand!"* The men in the second rank place one foot backward so as to get a steady posture, and put their arms up, the hands of the first rank resting in those of the second. Those giving the support should take care not to push the arms of the others forward (and to remain as rigid as possible). To avoid this tendency, the support may instead be given by one hand on the neck (Fig. 65). Still another way is to do the movement with neck firm, the support then given at the elbows (Fig. 66). This form is probably the least productive of faulty posture, as here the weight of the head presses the arms backward, forcing the chest to be vaulted. To use living supports is usually not suitable for children, although a skilled teacher may safely try it also for them.

Instead of *bending* backward, beginners are apt to *fall*

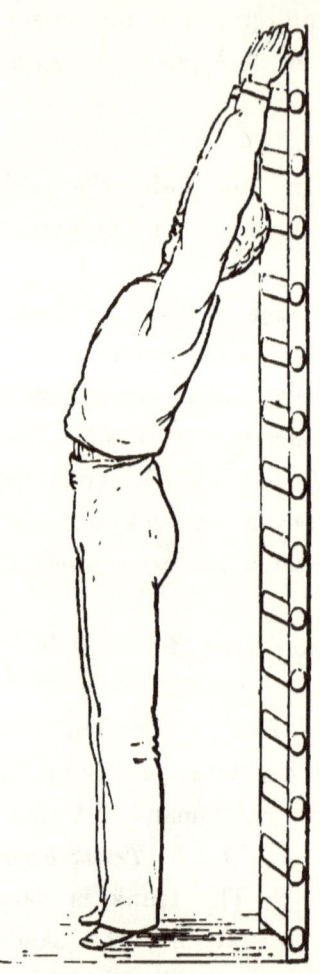

FIG. 61. — STR. GR. ARCH ST. POS.

MOVEMENTS OF THE SWEDISH SYSTEM 77

Fig. 62.—Half Str. Gr. Arch St.

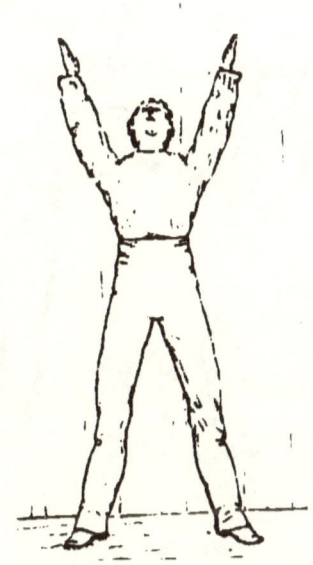

Fig. 63.

Fig. 64.

Fig. 65.

against the apparatus, thus compressing their chests and pushing their heads forward. To correct this, the movement should alternate with the following one:

Str. St. T. Backw. Flex. over Bar (Fig. 67).— The horizontal bar is put so low that it rests against

Fig. 66.

the sacrum, when the pupil stands close up to it with his back turned to it. Command. "*Arms upward— stretch! Trunk backward —bend!*" etc. The bar now compels the pupil to bend instead of falling backward. This movement can also be done with living supports (Fig. 68). Command, "*Numbers two (one) chain support— stand!*" The indicated numbers lock hands, and rest them in the small of the backs of those in front of them, and take

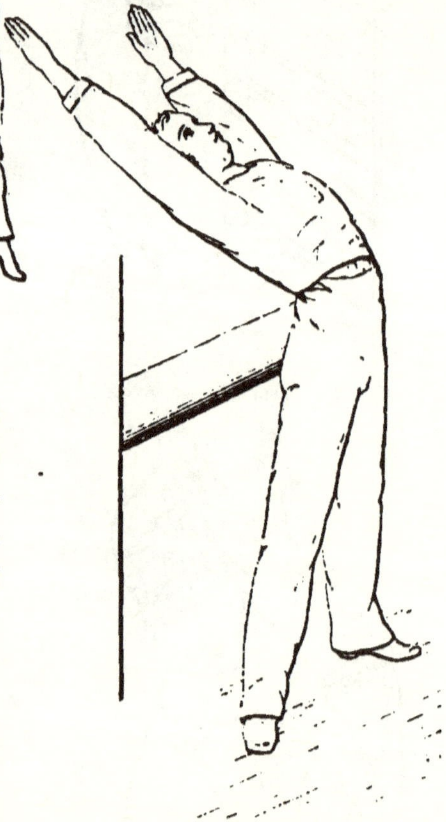

Fig. 67.— STR. STRIDE ST. T. BACKW. FLEX. OVER BAR.

the fallout st. *b* pos. forward, to gain a firm posture. The movement now takes place in usual order. Suitable only for grown persons.

All of the movements just mentioned are done in stride st. pos. as well, and the flexion over the bar can be made stronger

FIG. 68. — STR. ARCH SUP. ST. 2 A. EXT.

by adding 2 A. ext. in the arch st. pos.; the movement is then called

Str. Arch Sup. St. 2 A. Ext. (Fig. 68), and is executed as the corresponding free-standing movement. On the same principles we also have

Str. Arch Sup. Stride St. 2 A. Ext.

Str. Gr. Arch St. Heel-elev. (Fig. 69). — Command (after arranging the pupils at the apparatus). "*Arms upward —*

stretch! Trunk backward — bend! Grasp! Heels — lift! Sink! The same— óne! Two!" . . . Arms and legs must be kept straight. When the heels are raised, the chest becomes still more arched than when the feet rest fully on the ground. Unless the head is kept well backward and the body generally well balanced, it is not possible to do the movement without cramping the chest. The movement may be done with living supports (Figs. 64, 65, 66).

The same exercise can also be done

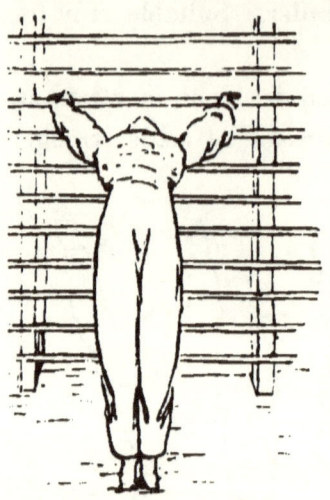

FIG. 69. — STR. GR. ARCH ST. HEEL-ELEV.

in stride pos., and in either position also with oblique grasp; thus we get the three exercises:

Gr. Arch Stride St. Heel-elev. (Fig. 70).

Oblique Gr. Arch St. Heel-elev.

Oblique Gr. Arch Stride St. Heel-elev. (Fig. 71).

Gr. Arch St. Kn. Flex. Upw. (Fig. 72). — After the commencing position (gr. arch st.) has been taken, command, "*Left (r.) knee upward — bend!*" The leg is bent so that the

FIG. 70. — GR. ARCH STRIDE ST. HEEL-ELEV.

thigh is horizontal, the lower leg vertical, and the foot pointing downward. The other leg remains straight. " *Change feet —*

one! Two!" . . . 1. The foot is put down. 2. The other leg is raised in a similar manner. The exercise is done also with oblique grasp. For advanced pupils, it may be done as follows: " *Knee-elevation — one! . . . Four!"* or, still later on, " *Knee-elevation — one! Two! One! Two!"* . . . when the leg is quickly raised as soon as the other foot touches the ground.

Gr. Arch St. Kn. Flex. and Ext. (Fig. 73). — Command, . . . " *Knee upward — bend! Knee — stretch! Bend! . . . Change

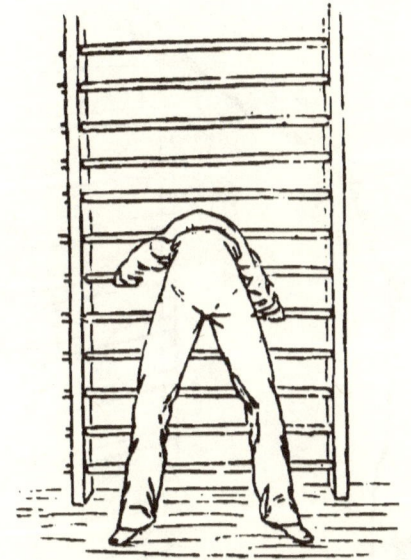

FIG. 71. — OBLIQUE GR. ARCH STRIDE ST. HEEL-ELEV.

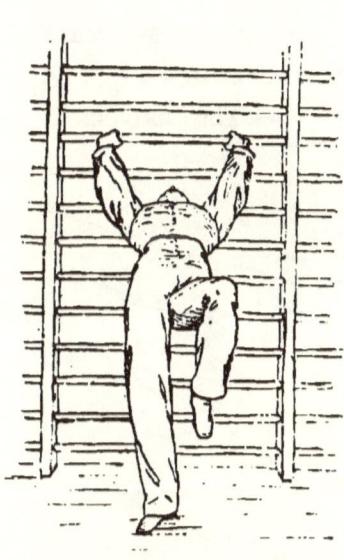

FIG. 72. — GR. ARCH ST. KN. FLEX. UPW.

feet — ône! Two!"* . . . The knee is stretched as in cr. half st., Kn. ext. forw. (see leg-movements). After some practice, the command will be, . . . " *Knee flexion and extension, left (r.) leg — one! . . . Four! Right (l.) leg — one! . . . Four!"* Done also with oblique grasp.

Gr. Arch St. L. Elev. — Command, . . . " *Left (r.) leg forward — lift!"* With straight knee and extended instep, the leg is raised as far as muscular resistance and good posture will allow. "*Change feet — one! Two! . . . Foot downward — place!*" This movement can be combined with heel-elevation.

Gr. Arch St. L. and Heel-elev. (Fig. 74). — Command, . . . "*Left (r.) leg forward — lift! Heel-elevation — óne! Two! . . Change feet!*" . . . Both of these exercises may also be done with oblique grasp.

Gr. Arch St. Hand Movement Downw. (Fig. 75). — The movement

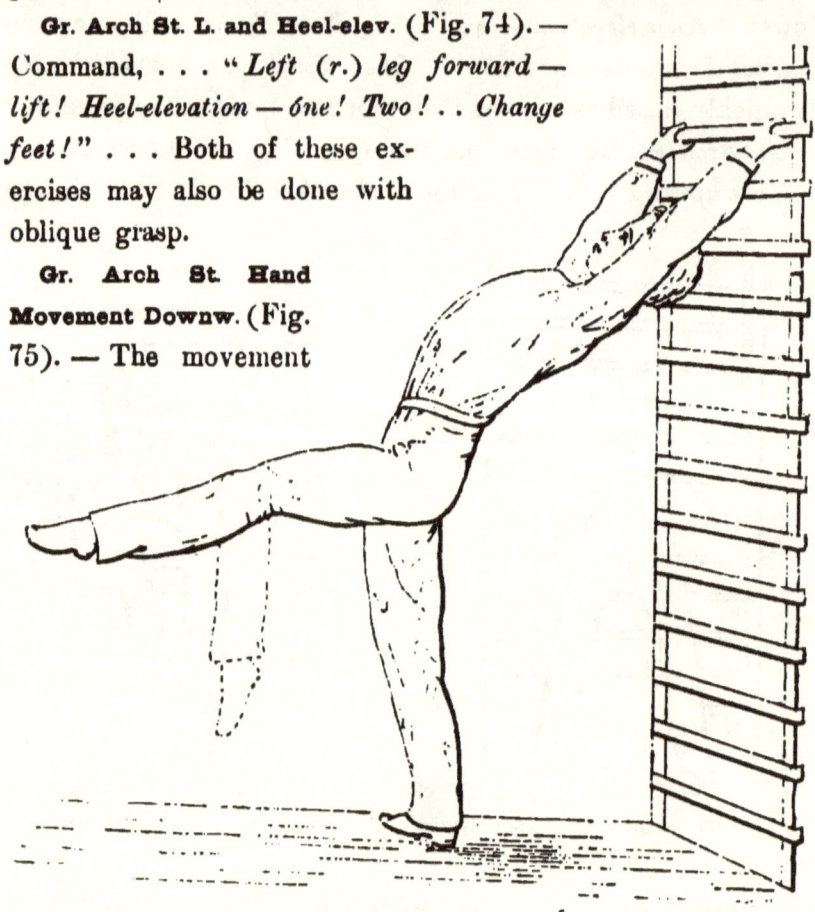

FIG. 73. — GR. ARCH ST. KN. FLEX. AND EXT.

can be done (1) so that the hands move to the same bar. Command, "*Hand to hand movement downward — one! . . . Four!*" 1. Left (r.) hand grasps the next lower bar. 2. The right (l.) hand is moved to the same bar. 3. The right (l.)

MOVEMENTS OF THE SWEDISH SYSTEM 83

hand is moved one bar down. 4. The left (r.) hand follows. "*Upward the same — one!* . . . *Four!*" takes place in the opposite order. (2) The hands skip one bar. Command, "*Hand

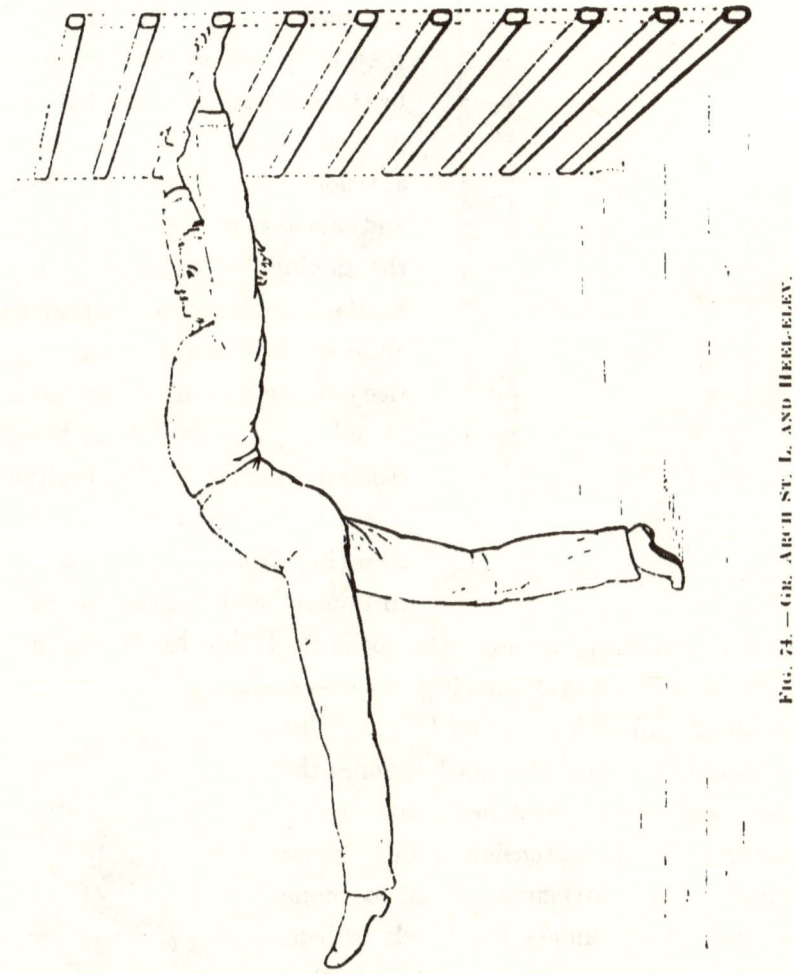

Fig. 74.—Gr. Arch St. L. and Heel-eley.

under hand movement — one! Two!" . . . 1. The left (r.) hand grasps the next lower bar. 2. The right (l.) hand grasps the bar below the one now held by the left (r.), thus skipping one bar, etc. (3) Both hands move simultaneously. Command, "*Double hand movement downward — óne! Twó!*" . . . The elbows are slightly bent; and by their rapid extension both hands are pushed from the bar, which they let go in order to grasp the next bar below (or next but one, if so ordered). The movements are done as well in stride pos., and always in the latter when the flexion becomes very deep. Limber individuals, especially children, will (after some practice) be able to move the hands from bar to bar until both hands rest on the floor. This position is called **gr. arch st. pos., hands on floor** (Fig. 76). A "backward contortionist" easily takes this position free-standing, by doing str. stride st. T. flex. backw. downw.

FIG. 75. — OBLIQUE GR. ARCH ST. POS.

Gr. Arch St. 2 A. Flex. (Fig. 77). — Command, . . . "*Arms — bend! Strétch! Bend!*" . . . The arms are bent until the head touches the bars, and again stretched, the flexion being slow, the extension a little more rapid. This movement cannot be done in good form unless the trunk is bent to horizontal position, or still farther backward (even until the hands rest on the floor). The same arm-movement is done in stride pos., as well as in the position shown in Fig. 74; and in any of these also with oblique grasp.

FIG. 76. — GR. ARCH ST. POS., HANDS ON FLOOR.

Stride Fall Hang. to Gr. Arch Stride St. Pos. (Fig. 78). — After the pupils have taken the fall hanging pos., with under grasp (for which see heaving-movements), command, "*Arch standing position — óne! Two!*" . . . 1. The feet, which previously rested on the heels, are now put fully on the floor, the body is pushed forward into gr. arch st. pos., the radial border of the hands resting on the bar, the thumbs held apart from the fingers, and their cushions firmly resting against the bar. 2. Commencing position is resumed by turning the hands

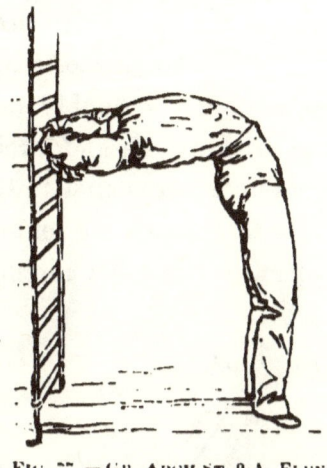

FIG. 77. — GR. ARCH ST. 2 A. FLEX.

over, raising the balls of the feet, and letting the body fall backward. This movement contains the intermediate positions of forward somersault over the bar (see jumping and vaulting), to which exercise it

FIG. 78. — STRIDE FALL HANG. TO GR. ARCH STRIDE ST. POS.

is an introduction. All arch-flexions are increased in strength, not only by changing the movement as above, but also by increasing the distance from the bars, and by making the backward flexion deeper. Every arch-flexion should be followed by a forward downward flexion, as this movement will counteract the evils of the accelerated heart-beat, which is caused by the excessive strain on the aorta in the backward flexion; for, when the body is bent thoroughly forward downward, the back sides of the legs are passively and forcibly extended, which gives this movement an effect similar to that of slow leg-movements.

Small children are seldom able to do a correct arch-flexion on apparatus (or other support for grasp), as they fall backward

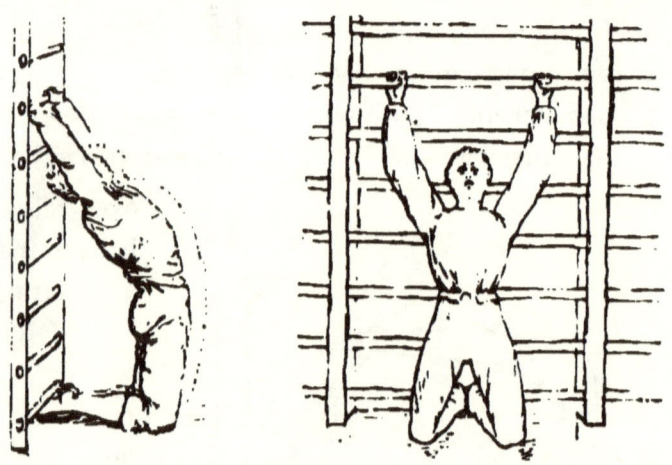

FIG. 79. — STR. GR. STRIDE KNEELING ARCH FLEX.

instead of bending. To correct this tendency, we give them a movement called

Str. Gr. Stride Kneeling Arch Flex. (Fig. 79). — The pupil kneels down in front of the stall-bars with his feet resting against the wall, and his hands grasping as high a bar as he can reach. Command, "*Arch flexion — one! Two!*" . . . 1. The feet are braced against the wall, and the body arched forward so that the knees rise from the floor. 2. Commencing position is resumed.

MOVEMENTS OF THE SWEDISH SYSTEM   87

Progressive arrangement of the arch-flexions: —

1. Wing St. T. Flex. Backw. and Forw.
2. Wing Stride St. T. Flex. Backw. and Forw.
3. Wing Close St. T. Flex. Backw. and Forw.
4. Wing Walk St. *b* T. Flex. Backw. and Forw.
5. Str. St. T. Flex. Backw. and Forw.
6. Str. Stride St. T. Flex. Backw. and Forw.
7. Str. Stride St. T. Flex. Backw., Forw., and Downw.
8. Str. St. T. Flex. Backw., Forw., and Downw.
9. Gr. Arch St. Pos.
10. Str. St. T. Flex. Backw. over bar.
11. Str. Walk St. *b* T. Flex. Backw. and Forw. Downw.
12. Bend Arch Sup. St. 2 A. Ext.
13. Gr. Arch. St. Heel-elev.
14. Gr. Arch Stride St. Heel-elev.
15. Str. Arch Stride St. 2 A. Ext.
16. Fall Hang. to Gr. Arch St.
17. Str. Arch Walk St. *b* 2 A. Ext.
18. Rest Stride St. T. Flex. Backw. and Forw.
19. Str. Arch Walk St. *b* 2 A. Ext. and Change of Feet.
20. Rest Arch St. Heel-elev. (living support).
21. Gr. Arch St. Kn. Flex. Upw.
22. Gr. Arch St. Kn. Flex. and Ext. Forw.
23. Gr. Arch St. L. elev.
24. Gr. Arch St. L. and Heel-elev.
25. Gr. Arch St. 2 A. Flex.
26. Gr. Arch Stride St. Hand to Hand Move.
27. Oblique Gr. Arch St. Heel-elev.
28. Oblique Gr. Arch Stride St. Heel-elev.
29. Gr. Arch St. L. and Heel-elev. with Deep Flexion.
30. Gr. Arch Stride St. 2 A. Flex.
31. Gr. Arch Stride St. Hand under Hand Move.
32. Oblique Gr. Arch St. L. elev.
33. Gr. Arch Stride St. 2 A. Flex. with Deep T. Flex.
34. Oblique Gr. Arch St. 2 A. Flex.
35. Oblique Gr. Arch St. L. and Heel-elev.
36. Oblique Gr. Arch Stride St. 2 A. Flex.
37. Gr. Arch Stride St. 2 Hand Move.
38. Oblique Gr. Arch St. L. and Heel-elev. with 2 A. Flex.
39. Gr. Arch St. Pos. w. Hands on the Floor.
40. Gr. Arch St. w. Hands on Floor and 2 A. Flex.

HEAVING-MOVEMENTS.

By heaving-movements, we mean exercises which expand the chest by drawing the superior ribs apart, besides developing the muscles of the arms; they consist of various forms of lifting the body by means of the arms alone, or by using both arms and legs. Before you hang by means of the arms alone, the extensibility of the front side of the body must have been cultivated by arch flexions, and that of the arms and lateral chest by double arm-extensions. For this reason, the last-named movements are to be considered as introductions to the heaving-movements, and as such will be given a place in the ensuing list.

The heaving-movements may commence as early as the second day's lesson, if any apparatus is at hand. After two or three months' practice, each lesson should contain two heaving-movements, one immediately after the arch flexions and another immediately before the jumping; then the two are made as different as possible, and the last one should be the stronger of the two. In total absence of apparatus, 2 A. ext. and a greater number of lateral trunk-movements and respiratory exercises will have to form the substitute.

FIG. 80. — ST. 2 A. FLEX. UPW. BEND ST. POS.

St. 2 A. Flex. Upw. (Fig. 80). — Command, "*Arms upward — bend!*" The upper arms remain still; with moderate speed the forearms are bent upward as far as possible with the hands semi-closed in front of the shoulders and brought as far sideways as possible. The elbows should be held close to the body. At the command, "*Position!*" the arms resume fundamental position. When the arms are bent upward, the position is called **bend st., bend sitt.**, etc., and this is the intermediate position of all arm-extensions.

2 A. Ext. Upw. (Fig. 81 *a*). — Command, "*Arms upward stretch — one! Twó!*" 1. Bend st. pos. 2. The arms are quickly extended upward, so that they are parallel, the palms of the hands turned toward each other,[1] the fingers of each hand held together, and the arms carried as far backward as possible. This position is called **stretch** (str.) **pos.** Common faults are to push the head and abdomen forward.

2 A. Ext. Forw. (Fig. 81 *b*).—Command, "*Arms forward stretch — one! Twó!*" 1. Bend st. pos. 2. The arms are thrust quickly forward; the arms are parallel and horizontal, and the palms of the hands turned toward each other. This position is called **reach st.** Common faults are to push the shoulders and abdomen forward and to bend the trunk backward.

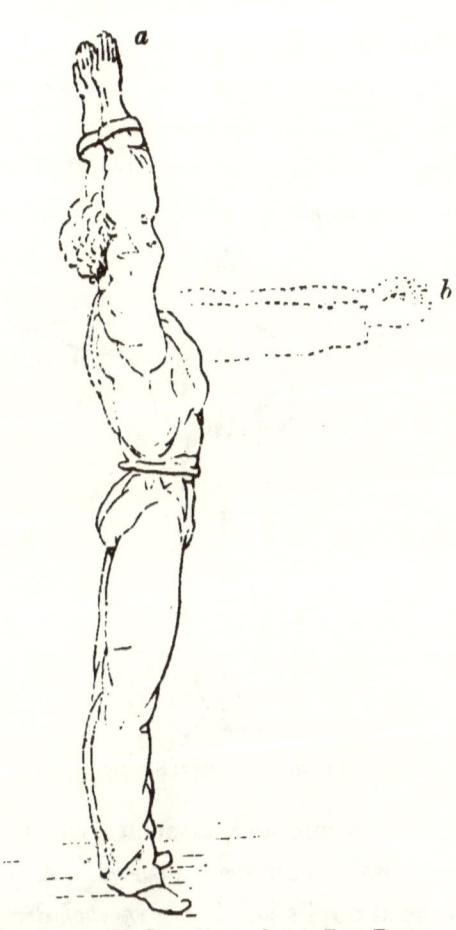

FIG. 81.—*a*, 2 A. Ext. Upw.; *b*, 2 A. Ext. Forw.

2 A. Ext. Sidew. (Fig. 82).—Command, "*Arms sideways stretch — one! Twó!*" 1. Bend st. pos. 2. The hands are thrust in a straight line sideways from the shoulders, so that

---

[1] If the palms are turned the opposite way, the chest becomes compressed and the head pushed forward.

the arms become straight, horizontal, and carried well backward, the palms of the hands being turned downward. This position is called **yard st.** *c* (for other yard st. positions, see shoulder-blade movements).

To resume fund. pos. from any one of these positions, command, "*Arms downward stretch — one! Twó!*" 1. Bend st. pos. 2. The arms are stretched down into fund. pos.

As the pupils gain proficiency, the command grows shorter; thus, if at first it is, "*Arms upward stretch — o n e! Twó!*" later on it will be, "*Arms upward — strétch! Twó!*" and still later on, "*Arms upward — strétch!*"

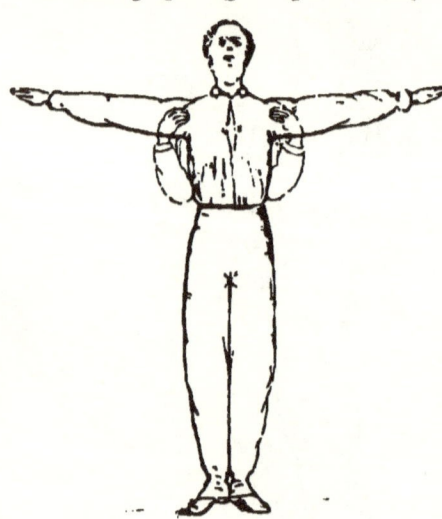

FIG. 82. — 2 A. EXT. SIDEW.

In absence of apparatus, these movements serve as substitutes for the heaving-movements, and may then (as well as at other times) be done twice or several times in each direction. Otherwise they are used as introductions, and as such form part of every lesson, even for the most advanced classes. A favorite combination for advanced pupils is, "*Twice in each direction, arms forward, upward, sideways, and downward — strétch!*" (or, "*stretch — one! Twó! One!*"...). For beginners, the reach st. pos. should be tried last, as it is the most productive of faulty posture; it should always be immediately followed by the yard st. pos., as this one has the quality of correcting these same faults.

When used as introductions, the 2 A. exts. may be combined with a leg-movement; as, for instance, the str. stride st. pos. (Fig. 83); command, "*Left (r.) foot sideways place and arms upward stretch — óne! Twó!*" 1. Bend st. pos. 2. The foot is moved, and, at the same time, the arms are stretched as previously described. (Compare leg-movements: stride st. pos. for children; the command is then, "*Feet sideways place and arms upward stretch — óne! Twó!*" each movement of the arms being accompanied by one foot-movement.) In this manner, the various walk st. positions can be practised, so that the pupils get some skill in taking the different commencing positions quickly and correctly. These introductions are excellent for the improvement of a child's conception of rhythm and co-ordination of motion. (For other arm-extensions, see shoulder-blade movements.)

When hanging by the arms alone from a horizontal bar, we can grasp the latter in three different ways, as follows:

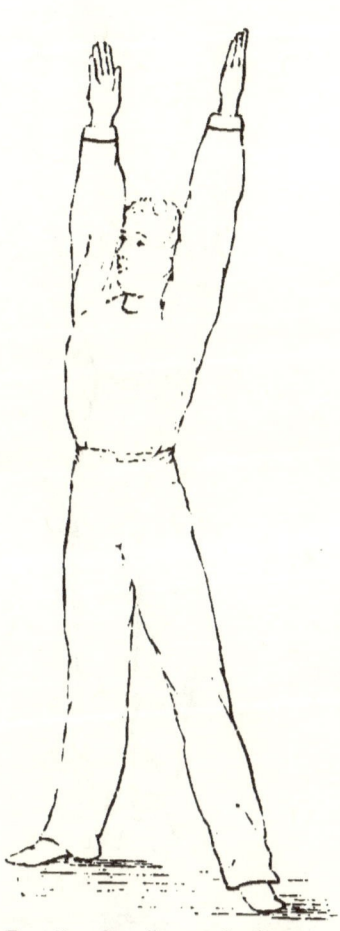

FIG. 83. — STR. STRIDE ST. POS.

Under Gr. Hang. Pos. (Fig. 85). — The pupils stand facing the bar. Command, "*Under grasp — grásp!*" The pupils jump, and grasp the bar on the opposite side from that on which they stand. The arms are straight, and the distance between the hands at least equal to the shoulder-width. If you wish the pupils to grasp on the same side on which they stand, the position is

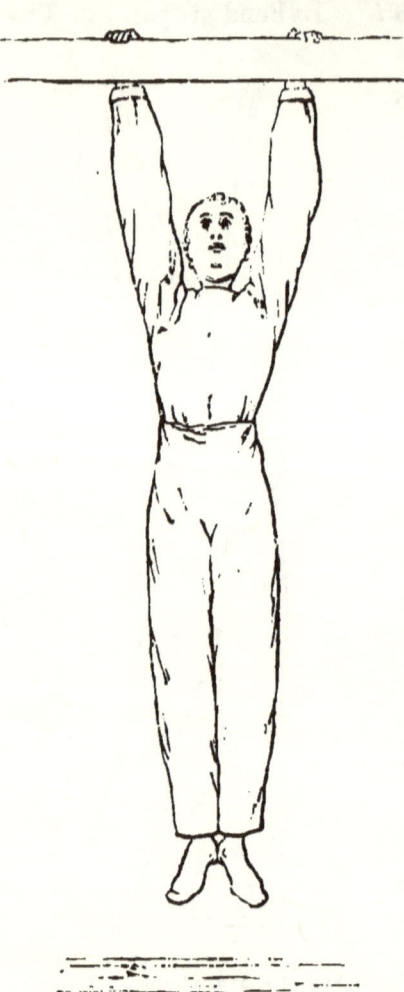

FIG. 84.—OVER GR. HANG. POS.

Over Gr. Hang. Pos. (Fig. 84).—Command, etc., on the same principles as the previous exercise. If one hand grasps each side of the bar, the position is called

Under Hang. Pos. (Fig. 86), which is done on the same principles. (Best with the groove up.)

When these positions are practised for the first time, the bar is put so low that every pupil can easily reach it, those who are very tall sometimes having to bend their knees slightly backward, in order not to stand on the floor after the hands have grasped.

In each one of these positions the pupils may pull themselves up by bending the arms, the exercises being:

Under Gr. Hang. 2 A. Flex. (Fig. 85).—Command (after commencing position is taken as above), "*Lift!*" With moderate speed, the pupils raise themselves as high as possible, the

MOVEMENTS OF THE SWEDISH SYSTEM

head, elbows, and (straight) legs being carried well backward and the chest forward. At the command "*Sink!*" the pupils lower themselves slowly until the arms are straight.

Over Gr. Hang. 2 A. Flex., and

Under Hang. 2 A. Flex. are done on the same principles. These movements may be increased in strength by increasing the distance between the hands, by bending the arms only to 90°, or by increasing the duration of the movement (by repeating it a number of times). When repeated several times, the movement should not be done by quick jerks, but slowly and evenly, a pause being made in each of the two positions (bend gr. hang. and str. gr. hang. positions).

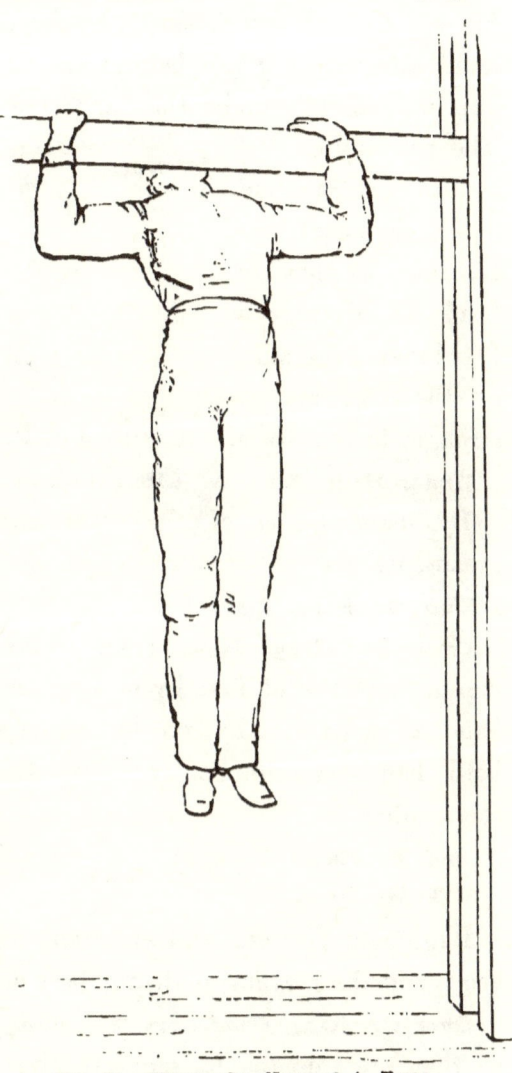

FIG. 85. — UNDER GR. HANG. 2 A. FLEX.

Next in progression comes travelling on the bar. This can be done in either of the above positions with the arms bent.

Under Hang. Travel. (Fig. 86).—Command, "*Stárt!*" The pupil stands at one end of the bar, grasps it as above, pulls himself up, and travels slowly to the other end by letting each hand alternately grasp behind the other one, the head being kept immediately under the bar ("the nose in the groove!"). The bar may be horizontal or inclined. The movement can be done so that the head is alternately lifted above the sides of the bar, and is stronger in that form. It is then called

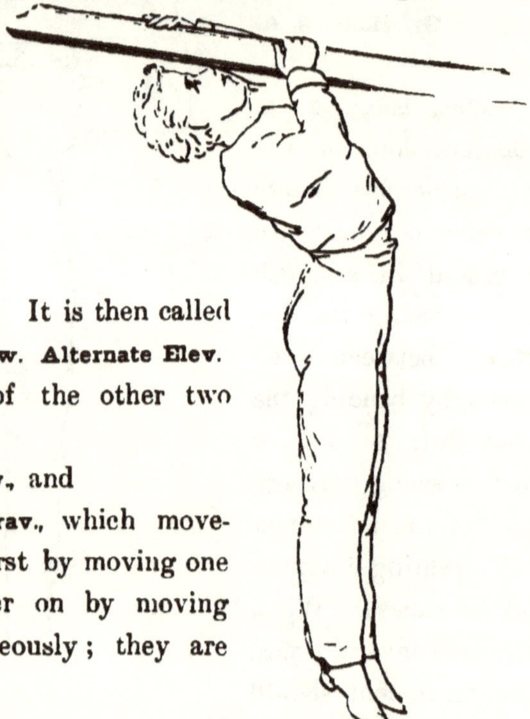

Under Hang. Trav. w. Alternate Elev.

By using either of the other two grasps, we get

Und. Gr. Hang. Trav., and

Over Gr. Hang. Trav., which movements are done at first by moving one hand at a time, later on by moving both hands simultaneously; they are then called

Und. Gr. Hang.
Over Gr. Hang. } **2 Hand. Trav.**

FIG. 86.—UNDER HANG. 2 A. FLEX., OR UNDER HANG. TRAV.

Beginners (women and children) may travel with straight arms, and the movement then is best done as

Over Gr. Hang. Oscillatory Travelling, which exercise is done so that the impetus for motion is received by oscillating the whole body (from the shoulders to the feet) from side to side, the left hand moving when the legs swing to the left; the right, when they swing to the right. This exercise corresponds to travelling in the rings (common in American gymnasiums), but is safer and more developing than the latter.

One of the strongest horizontal-bar movements is the
Rotary Travelling (Fig. 87). — This movement is done as follows: Grasp the bar with the left (r.) hand, and hang with this arm bent (as shown in Fig. 87 a), the right (l.) arm hanging down by the side; the head is on the side of the bar opposite the hand that grasps. Now tip the head backward under the bar, and rotate the body to the right (l.) until the head is on the same side of the bar as the left (r.) hand; now grasp with the right (l.) hand, the right (l.) arm bent at the elbow, and let go with the left (r.) hand.

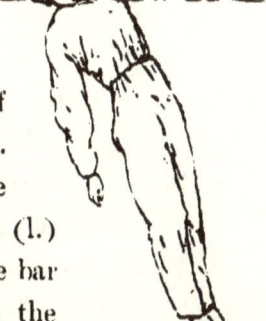

Fig. 87a.—Rotary Travelling on the Horizontal Bar.

Hanging on the bent right (l.) arm, you rotate forward to the right (l.) so that the head comes on the original side of the bar, when again grasp with the left (r.) hand, and continue the movement as just described until the other end of the bar

Fig. 87 b. — Rotary Travelling on the Horizontal Bar.

is reached. Fig. 87 b shows the grasp of the hands, the movement starting with both hands on the bar. If the bar is grasped so that the first rotation takes place forward, it is a little easier to continue the movement.

Travelling by means of the arms alone can also be done upward between two ropes.

Under Hang. Vertical Trav. (hand above hand) (Fig. 88). — Grasping two of the vertical ropes, one in each hand, the pupil pulls himself up as far as possible, extends one arm upward, and grasps the rope a little higher, bends that arm to utmost flexion, and repeats the movement with the other arm, and with each hand alternately until he reaches the ceiling. He travels downward in the same manner or by moving both hands at the same time, — a short distance at a time, and with the arms always bent. Before any upward travelling takes place, 2 A. flex. alone (Fig. 89) should be practised; and for advanced pupils this movement may be done two or three times after the ceiling has been reached (or when the pupil has gone as high as he can) before travelling down again.

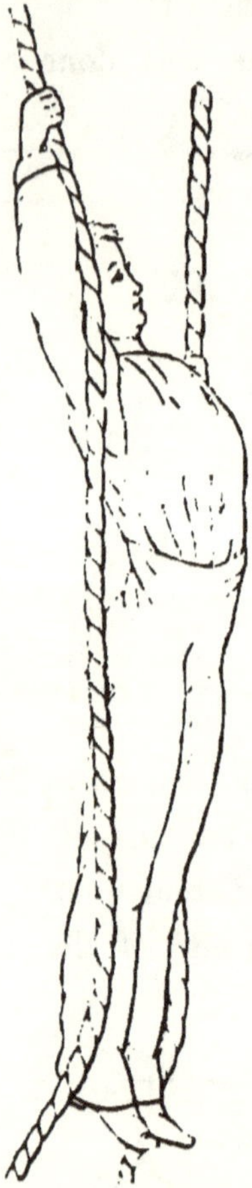

Figs. 90, 91, and 92 show how a corresponding exercise can be done on ladders and stall-bars, the movement then being done with over grasp.

FIG. 88.—UNDER HANG. VERTICAL TRAV.

FIG. 89. HANG. 2 A. FLEX.

Between the ropes, we can also turn a somersault backward, if we hang on straight arms, lift the legs forward upward, and tip the head and shoulders backward, coming back the opposite way. This is called

Somersault Backw. betw. Ropes (Fig. 93). — In this move-

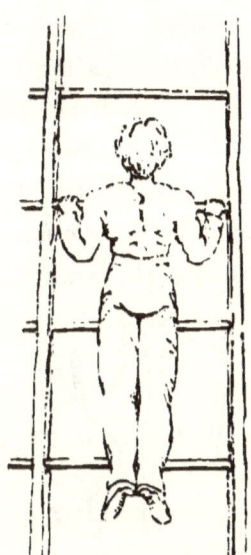

FIG. 90. — VERTICAL TRAVELLING ON PERPENDICULAR LADDER.

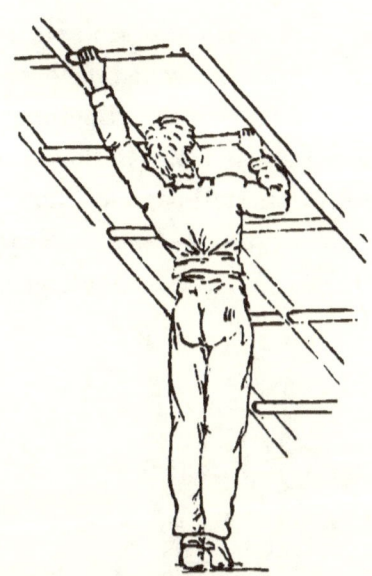

FIG. 91. — OVER GR. HANG. TRAV. ON INCLINED LADDER.

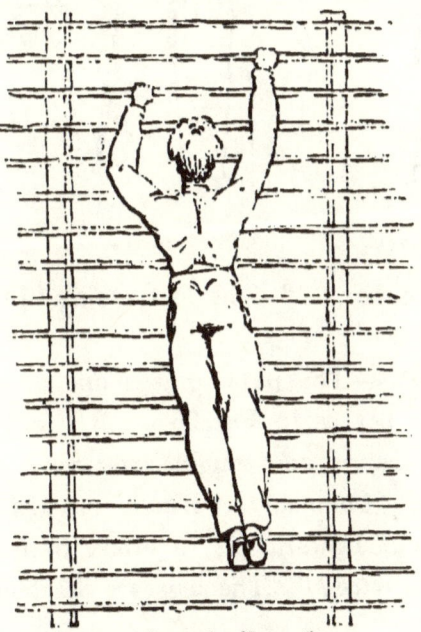

FIG. 92. — OVER GR. HANG. TRAV. ON STALL-BARS.

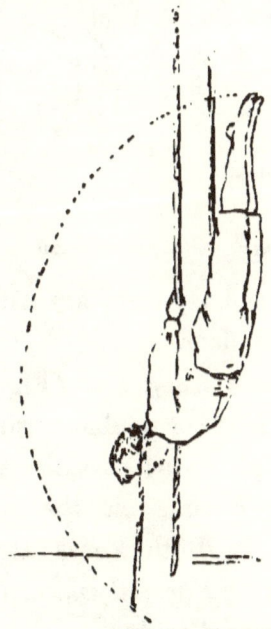

FIG. 93. — SOMERSAULT BACKWARD BETWEEN ROPES.

ment, skilled athletes are able to hold themselves in horizontal position with upturned face or facing downward (after turning the somersault). Beginners should be assisted as shown in Fig. 94 *a* and *b*. Children may be given the corresponding exercises on stall-bars, as shown in Fig. 95. Command, "*Forward downward — bend! Grásp! Stoop hanging position — one! Two!*" . . . 1. The legs are lifted so that the whole back

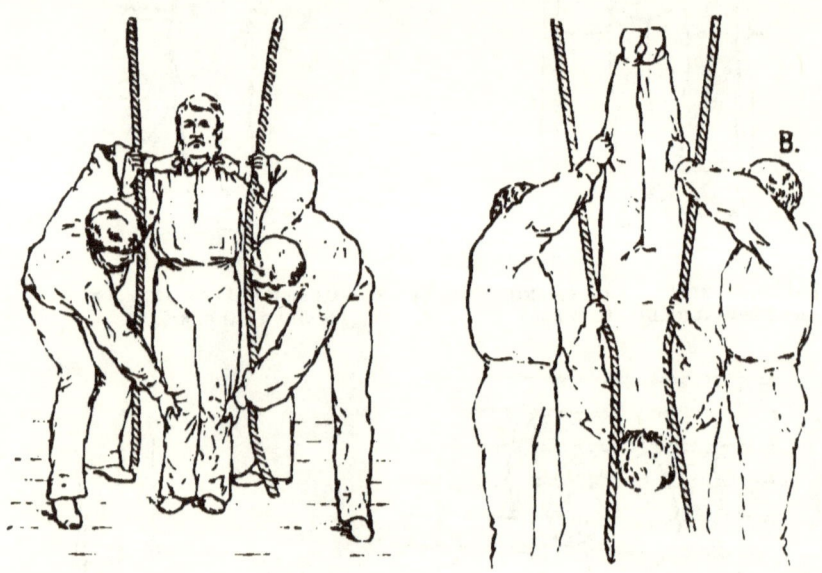

FIG. 94 *a*. — SOMERSAULT WITH ASSISTANCE.   FIG. 94 *b*.

of the body rests against the bars. 2. The feet are again put on the floor.

Crook Hang. Pos. (Fig. 96). — The pupil grasps one of the stall-bars (or double or single bar, as in Fig. 96 *b* and Fig. 97) as high up as possible, and hangs on straight arms with his back resting on the bars. Command, "*Knees upward — bend!*" Both knees are bent upward (as in crook half st. pos.) "*Knees downward — stretch!*" The legs are stretched downward again.

MOVEMENTS OF THE SWEDISH SYSTEM 99

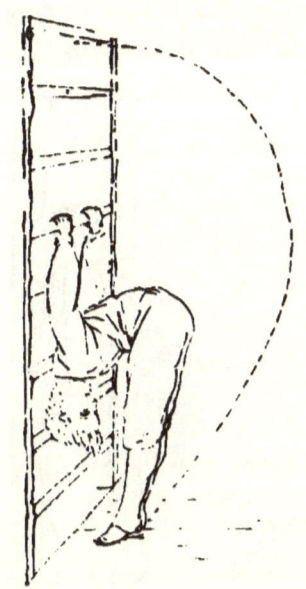

FIG. 95. — STOOP HANG. POS. ON STALL-BARS.

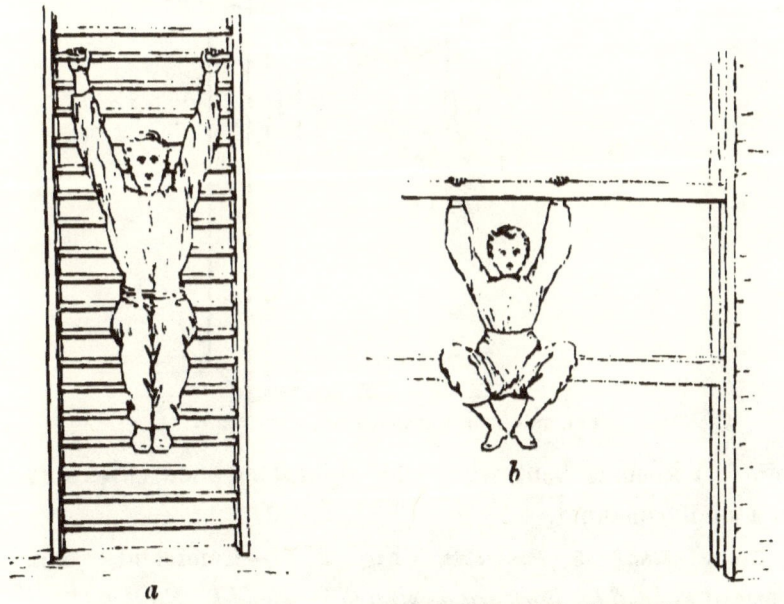

FIG. 96. — CROOK HANG. POS. *a*, ON STALL-BARS; *b*, ON DOUBLE BAR.

Crook Hang. Alternate Kn. Ext. — Command, "*Knees upward
— bend! Knee extension — óne! Twó!*" . . . 1. The left (r.)
knee is extended forward so that the leg is horizontal. 2. The

FIG. 97. — CROOK HANG. 2 KN. EXT. FORW.

left (r.) knee is bent while the right (l.) knee is extended
in a similar manner.

Crook Hang. 2 Kn. Ext. (Fig. 97). — Command, "*Knees
upward — bend! Forward — stretch! Bend! Stretch!*" . . .
Both legs are stretched simultaneously.

MOVEMENTS OF THE SWEDISH SYSTEM 101

Crook Hang. 2 L. Elev. — Command, "*Grásp! Legs forward — lift!*" Both legs are lifted to horizontal position. "*Sink!*" The legs are lowered.

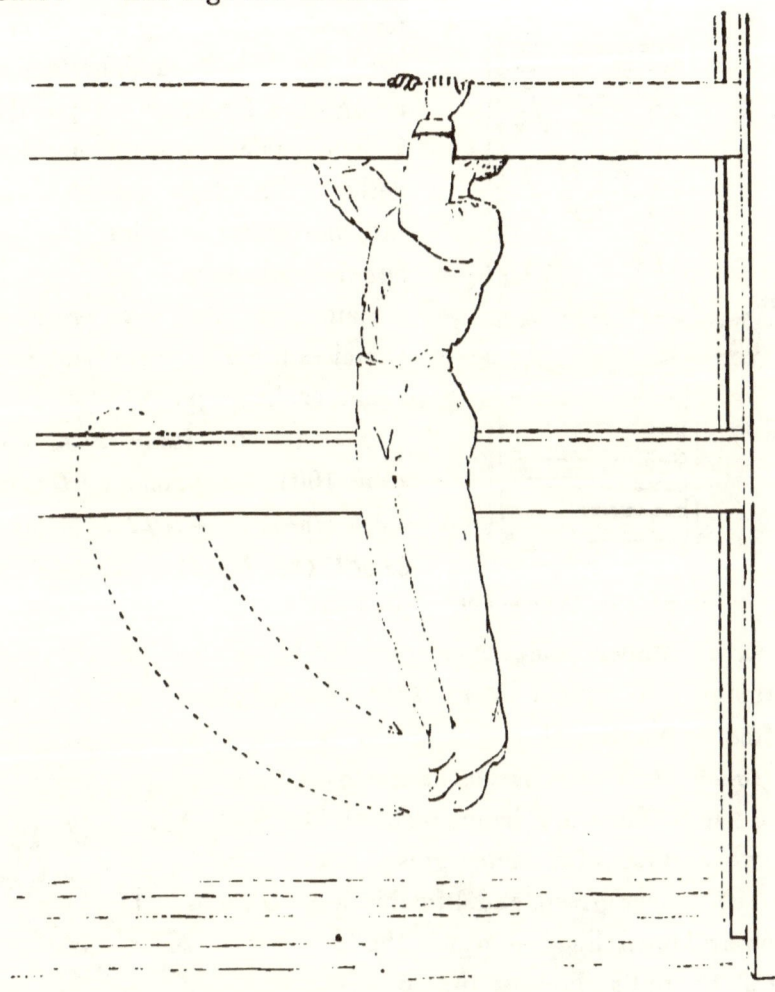

FIG. 98. — UNDER HANG. 2 L. SWING. OVER BAR.

Under Hang. 2 L. Swing. over Bar (Fig. 98). — Apparatus: double-bar. Command, "*Grásp!*" The pupil grasps the upper bar. "*Legs forward swing — one! Two!*" ... The pupil swings his legs straight forward over the bar from side to side.

Crook Hang. L. Abd. (Fig. 99). — Command, "*Grásp! Legs forward — lift! Leg abduction — one! Two!*" . . . 1. The legs are brought as far apart as possible. 2. The legs are again brought together. The legs remain lifted into horizontal position throughout the movement.

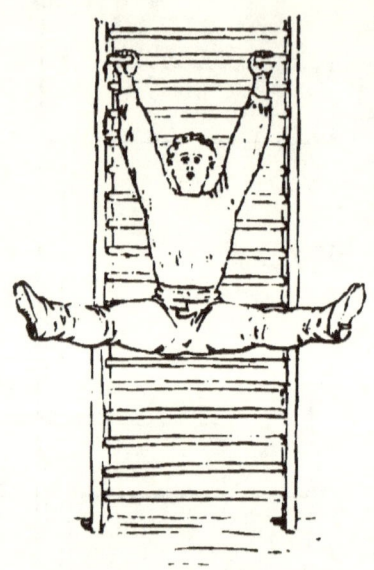

FIG. 99. — CROOK HANG. L. ABD.

Later on, the six preceding exercises are done with bent arms (bend cr. hang. pos.).

Over Gr. Cr. Hang. 2 A. Flex (Fig. 100). — Command, "*Grásp! Legs forward — lift! Arms — bend!* (or *Lift!*) *Stretch!* (or *Sink!*)" . . .

Crook Under Hang. Trav. (Fig. 101) is done on the same principles as under hang. trav., except that the legs are lifted to horizontal position.

Crook Hang. Somersault Backw. over the Bar. — The pupil hangs on straight arms, at first with under grasp, later on with over grasp, and lifts his legs forward until his feet touch the bar; now he pulls himself up on bent arms, and, in so doing, throws his legs over the bar (by tipping his head backward), tips his trunk backward (Fig. 102), as in the somersault between the ropes, and finally comes wholly

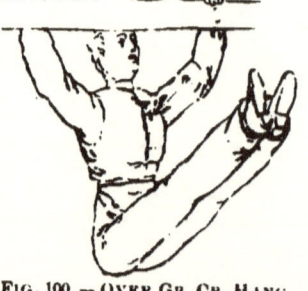

FIG. 100. — OVER GR. CR. HANG. 2 A. FLEX.

over on the other side of the bar on which he rests by his hands; and in this position the arms are straight, the abdomen leans against the bar, the body is gracefully curved backward, and the head is carried very high. This position is called

Balance Hang. Pos. (Fig. 103). — Another way of taking this position is to grasp the bar with both hands (over gr.), and with one knee over the bar, to swing the other leg quickly backward, so that you come up sitting astride the bar;

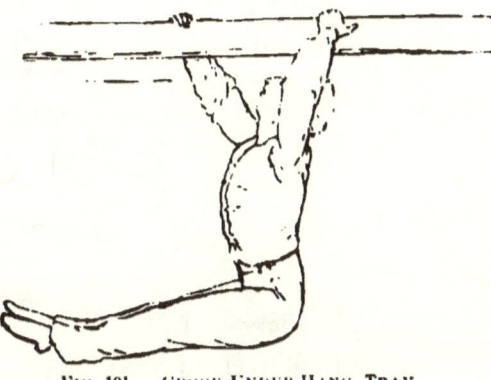

FIG. 101. — CROOK UNDER HANG. TRAV.

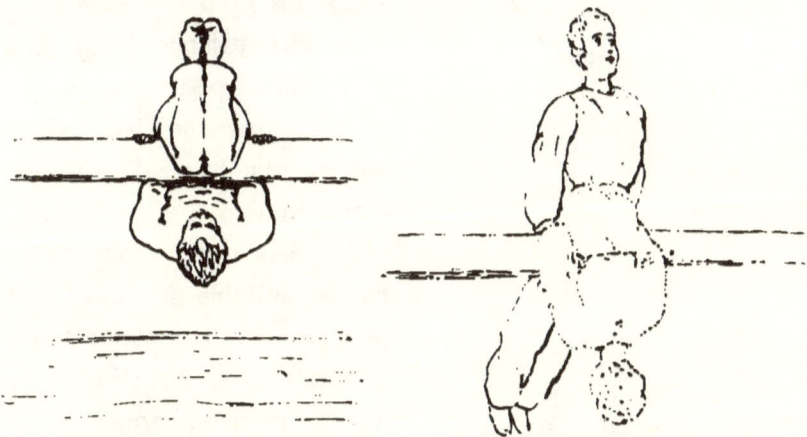

FIG. 102. — INTERMEDIATE POS. OF SOMERSAULT OVER THE BAR.

FIG. 103. — BALANCE HANG. POS.

then the forward leg is lifted backward over the bar and placed beside the other one. This movement is familiar to all "horizontal-bar performers," and hardly needs illustration. Instead of taking the balance hang. pos., you may bring the backward leg over the bar and sit on the latter (high sitt. pos.). From this position you can now turn a somersault backward, as shown in Fig. 104.

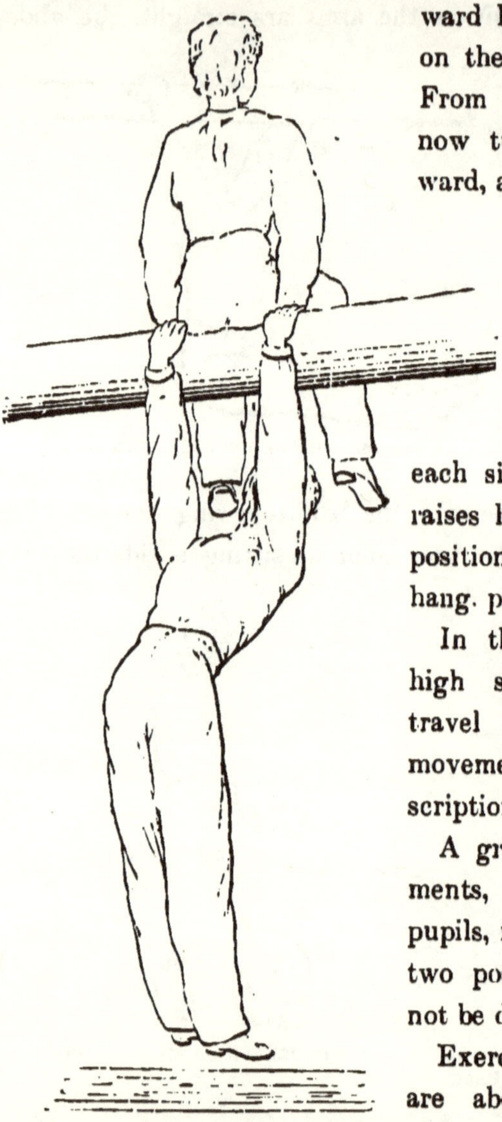

(In a schoolroom, the balance hang. pos. can be taken by the pupil standing in the aisle and placing his hands on the desks each side of him; if now he raises his knees forward, the position corresponds to the cr. hang. pos.)

In the balance hang. and high sitt. pos., we might travel along the bar, which movements need no description.

A great many other movements, suitable for advanced pupils, might be done in these two positions, but they will not be described here.

Exercises in cr. hang. pos. are abdominal exercises, as well; and, when a heaving movement of this kind is used, other abdominal exercises may be eliminated from that lesson.

FIG. 104. — HIGH SITT. SOMERSAULT BACKW.

Fall Hang. Pos. (Fig. 105). — The bar is put at hip-height. Command, "*Grásp!*" The pupils place their hands on the bar. "*Feet forward — pláce!*" The pupils swing their legs quickly forward under the bar, so that they hang on straight arms with heels resting on the floor. The body is straight from the head to the feet, or slightly arched, the head is backward, and the chest is vaulted. "*Position!*" The pupils resume commencing position. ("*The same — óne! Twó!*")

Arch Hang. Pos. (Fig. 106). — The bar is nearly shoulder-

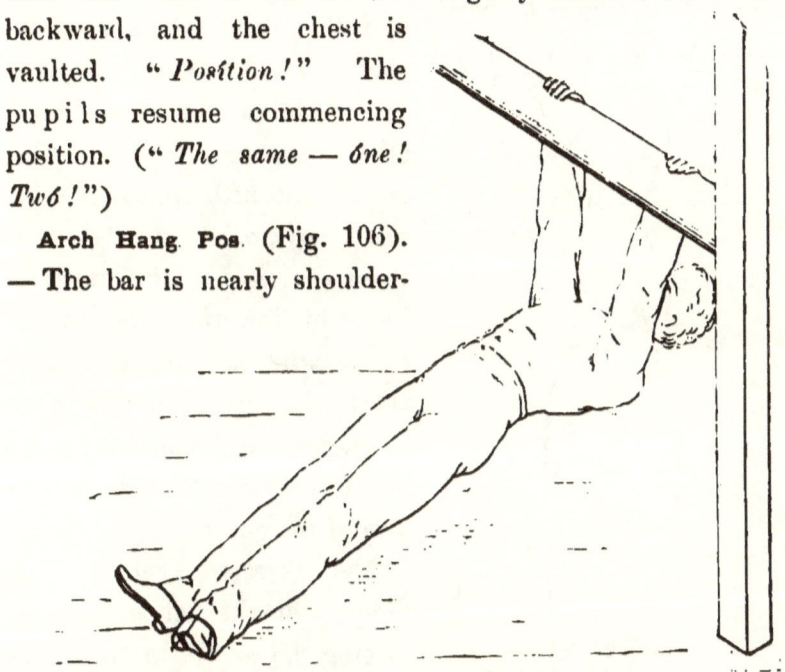

FIG. 105. — FALL. HANG. POS.

high. Command, "*Grásp! Feet backward — pláce!*" The feet are placed as far backward as possible, with the insteps stretched so that the body rests on the tips of the toes. The arms are straight. "*Position!*" etc., as above.

Change between Fall and Arch Hang. Pos. — Command, "*Feet forward — pláce! Feet backward — pláce!*" The arms remain straight while the feet are swung backward into arch hang. pos. "*The same — óne! Twó!*" . . .

In each position, we can do an arm-flexion on the same principles as in the other hanging positions. These exercises are called

Fall Hang. 2 A. Flex. and
Arch Hang. 2 A. Flex., and do not need any description.

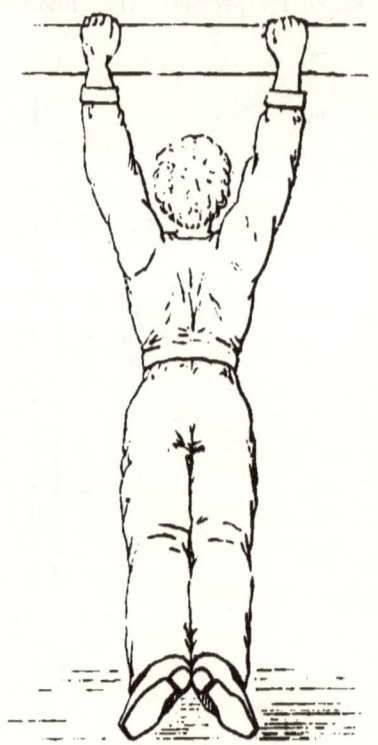

FIG. 106.— ARCH HANG. POS.

Of these five movements, the first, third, and fourth are suitable for children and beginners, as introductions to the various other hanging positions. The second and fifth are suitable for more advanced pupils. All these exercises are increased in force by lowering the bar or by bringing the hands farther apart. In this manner, the arch hang. pos. can be made to bring about the most forcible expansion of the chest.

The exercises can be arranged in the schoolroom by placing the pupils in the aisles (facing the back part of the room) and letting them grasp the desks on each side of them.

The following movements are various forms of climbing, which are especially suitable for children, although they should be done by grown persons as well.

Climbing the Inclined Rope (Fig. 107). — The hands grasp the rope, right (l.) hand above, and the left (r.) knee is swung over the rope, while the right (l.) leg hangs straight down. Travelling upward is now done in the following order: The left (r.) hand grasps above the right (l.) one, and at the same time the left (r.) leg is swung off the rope, and the right (l.) one swung on to it. The same move- ment is now done with the other hand and leg, and so on till the ceiling is reached. The movement can also be done on the horizontal bar. Before travelling is attempted, the movement of the legs should be practised alone, while one hand alternately grasps over and under the other one, which

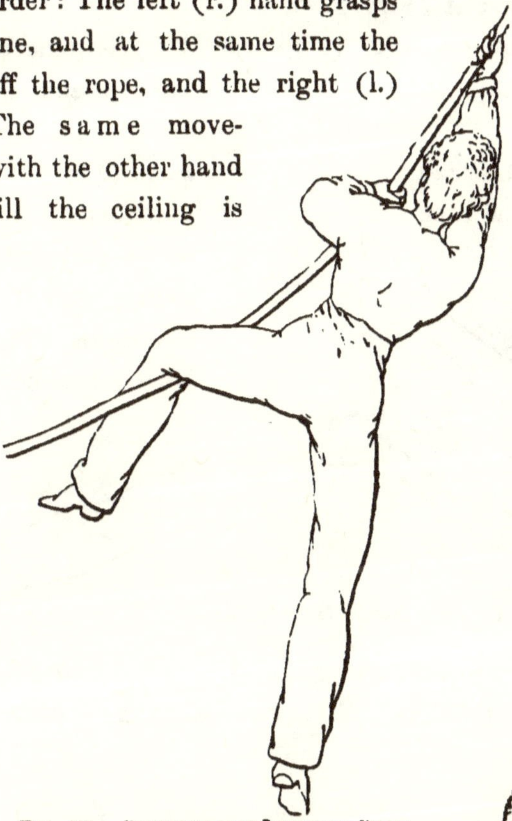

FIG. 107.—CLIMBING THE INCLINED ROPE.

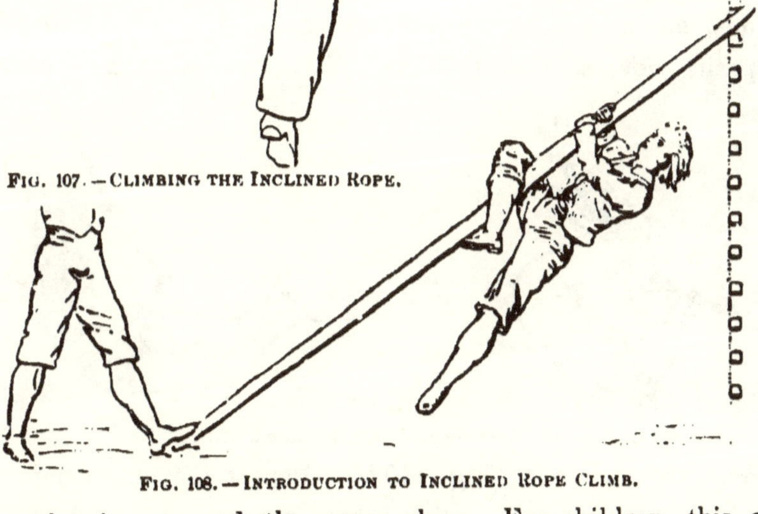

FIG. 108.—INTRODUCTION TO INCLINED ROPE CLIMB.

remains in one and the same place. For children, this can be arranged by tilting a long bar against the bar-stalls (or into a corner), while another pupil fixes its lower end by placing his foot against it (Fig. 108).

Inclined Rope Climb. Backw. (Fig. 109) is done on similar principles, except that the legs are up and the head down.

Rotary Climb. on Inclined Rope (Fig. 110). — Both hands grasp the rope with over grasp, the body is pulled up on bent arms, and both legs are swung over the rope above the hands (Fig. 110 a). Now move both

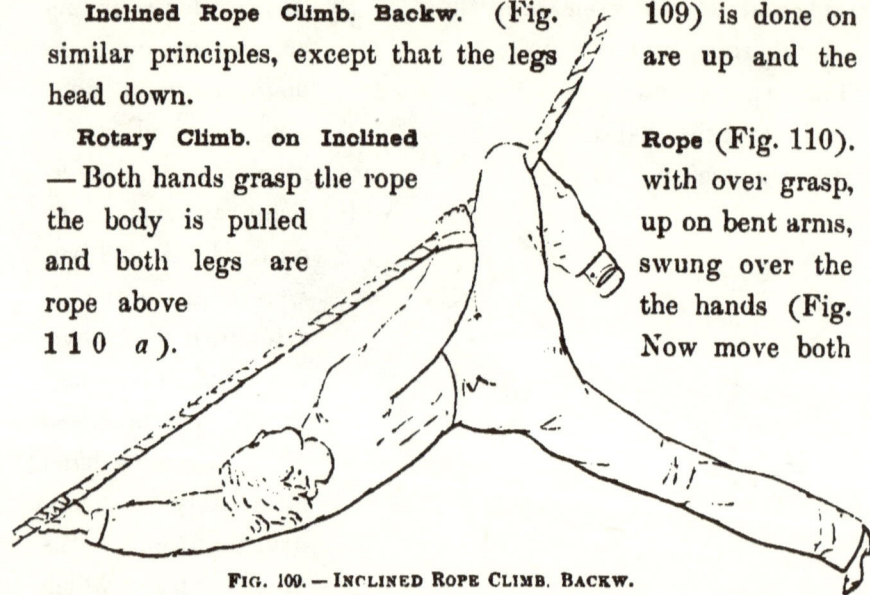

FIG. 109. — INCLINED ROPE CLIMB. BACKW.

hands, one at a time, upward, and grasp the rope from the opposite side, so that the hand that was below now comes

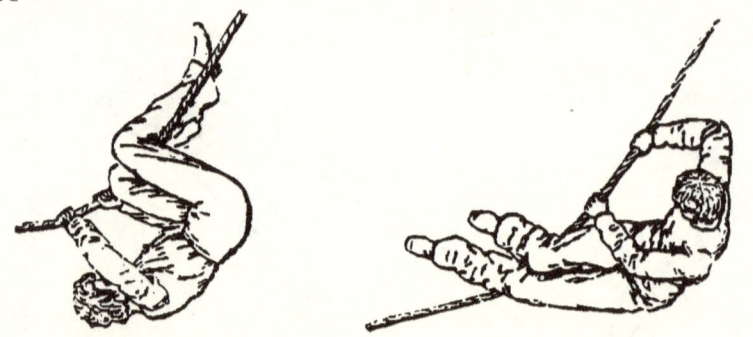

FIG. 110 a. ROTARY CLIMBING ON INCLINED ROPE. FIG. 110 b.

above the other one (Fig. 110 b). Repeat the movements of the legs, as above, and continue these procedures as high up on the rope as you can. (Suitable only for strong individuals.)

MOVEMENTS OF THE SWEDISH SYSTEM   109

In each of these movements you may come back by climbing downward on the same principles as you climb up in the first of the exercises. If possible, each pupil, upon reaching the ceiling, should climb down on a rope suspended from the same place as the inclined rope (or on some other apparatus), as this will save time.

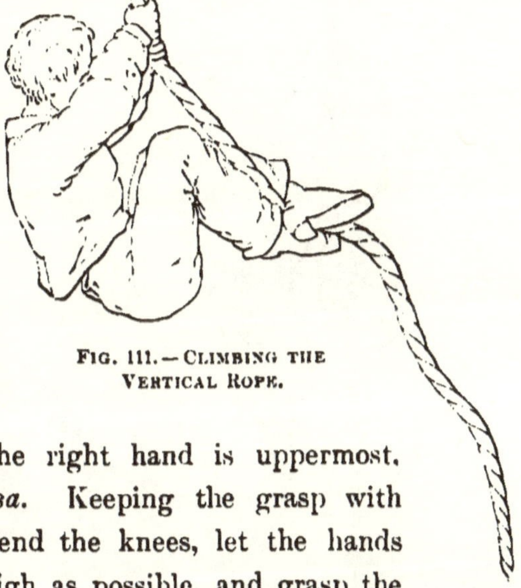

Vertical Rope Climb. (Fig. 111).— Grasp the rope with the hands, pull the legs up, and take the rope between the feet as near the hands as possible, the left foot being in front when the right hand is uppermost, and *vice versa*. Keeping the grasp with the feet, extend the knees, let the hands glide up as high as possible, and grasp the rope so that the hand which was below shall now be above the other one. Pull the legs up, and take a new grasp (as above), etc., all the way to the top. To come down, move hand under hand and let the feet slide down the rope, the legs being well drawn up. The movement can be done on the vertical poles as well (Fig. 112).

FIG. 111.— CLIMBING THE VERTICAL ROPE.

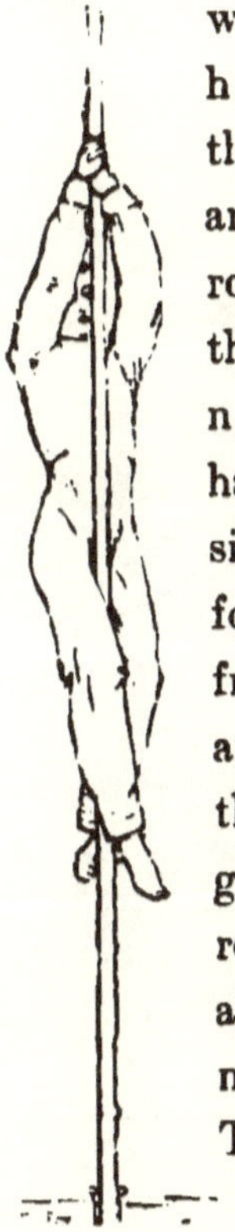

FIG. 112.—CLIMBING THE POLE.

Instead of coming down as just described, you can descend head first, hand under hand, as in Fig. 113, or by letting the feet slide on the rope, while the arms are extended sideways, as in Fig. 114. These modes, however, should not be attempted by any but advanced pupils, who, with assistance, have learned how to turn their feet up and grasp the rope, this being done near the floor.

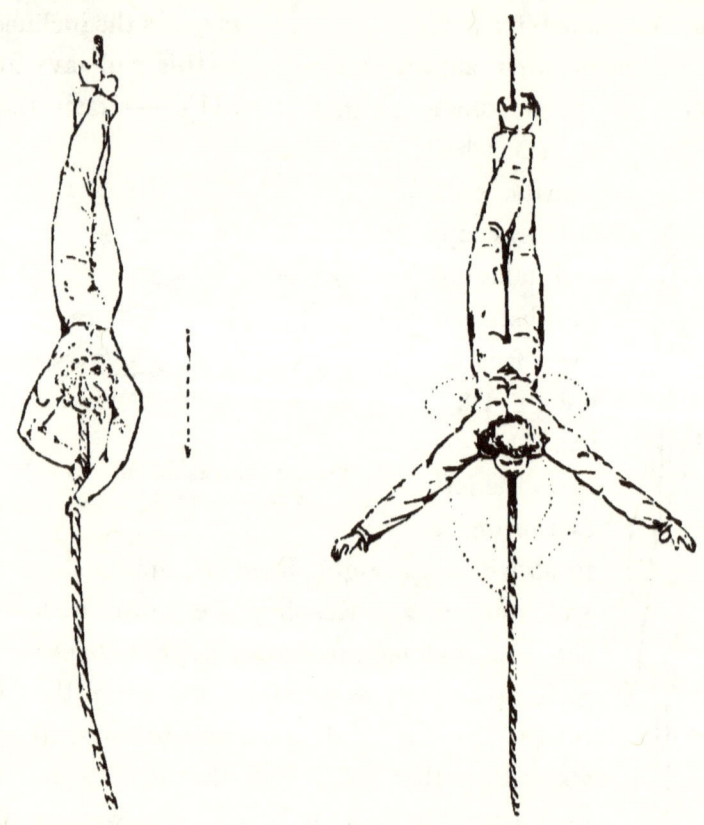

Fig. 113.—Climbing Down the Rope, Head First.   Fig. 114.—Sliding Down the Rope, Head First.

Trav. from Rope to Rope (Fig. 115). If several ropes (or poles) are hanging in a row, the pupils may travel from rope to rope, taking one grasp (and one "lift") on each rope, and steadily rising. When the ceiling is reached, they travel along on the same level till the last rope is reached, and then slide

MOVEMENTS OF THE SWEDISH SYSTEM 111

down in any of the ways described above. (Compare under gr. hang. vertical trav. on ropes.)

Serpentine through Vertical Ladder. — A pupil crawls into

FIG. 115.—TRAVELLING FROM ROPE TO ROPE.

one of the openings, and, sitting on one of the rounds, he grasps the next (or second) round above with his hands crossed; he pulls himself up into the next opening, helping himself by pushing with the feet (Fig. 116 a). He now faces the opposite way; and, grasping a round above, as before, he proceeds upward. When he has reached the top of the ladder, he moves into the next opening obliquely below, and climbs down, feet first, in a manner similar to his ascending (Fig. 116 *b*). As soon as there is room, another pupil starts; and in this way a constant current of pupils is kept going up in one half of the ladder and down in the other half.

FIG. 116. — SERPENTINE THROUGH VERTICAL LADDER.

Another way of arranging the movement is seen in Fig. 117, where the pupil climbs up zigzag and down the same way. This is usually done so that two pupils sit on the lowest round facing in opposite directions, start at the same time, and follow each other all the way up and down, crawling not only in and out through the openings, but, at the same time also, around each other.

MOVEMENTS OF THE SWEDISH SYSTEM 113

FIG. 118. — SERPENTINE AROUND THE SIDEPIECE OF LADDER.

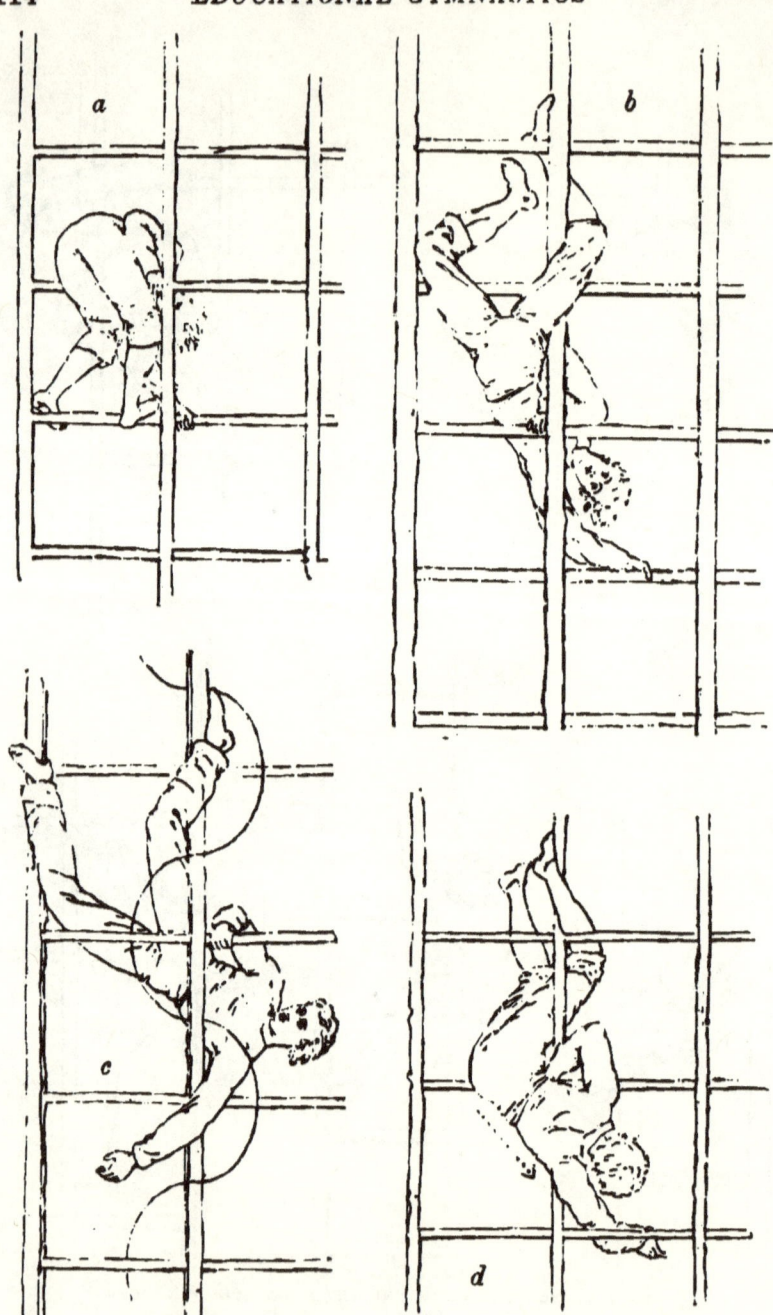

FIG. 119.—ZIGZAG SERPENTINE DOWNWARD, HEAD FIRST.

MOVEMENTS OF THE SWEDISH SYSTEM 115

The serpentine can also be done around the side-piece of the ladder, as shown in Fig. 118; *a* and *b* showing the movement

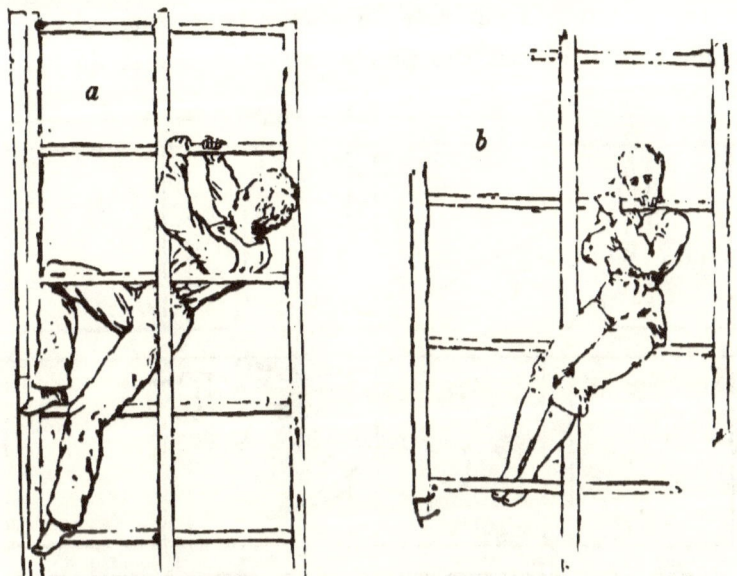

FIG. 117.—ZIGZAG SERPENTINE THROUGH VERTICAL LADDER.

upward; *c*, *d*, and *e* downward. You can also climb downward, head first, as shown in Figs. 119 and 120; 119 *a* showing the mode of turning: *b*, *c*, and *d* different stages of the movement.

Of these movements, the second is the easiest, but the first is used more, as it takes less time, if the class is large.

The corresponding form of climbing can be done in the horizontal ladder, the motion being similar to that described in the second of the preceding movements. The climbing is done either zigzag, head first upward and feet or head first downward (Fig. 121), or in and out through openings on the same level, either head first (Fig. 122), or feet

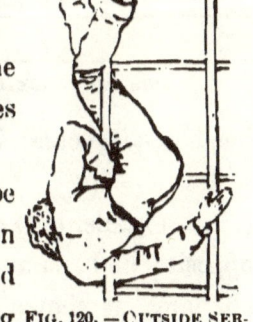

FIG. 120.—OUTSIDE SERPENTINE DOWNWARD, HEAD FIRST.

116 EDUCATIONAL GYMNASTICS

first (Fig. 123). In total absence of ladders, a similar movement is done on the double bar, as shown in Fig. 122 b, where the pupils climb from side to side around each other. In the horizontal ladder, the pupils can soon travel downward,

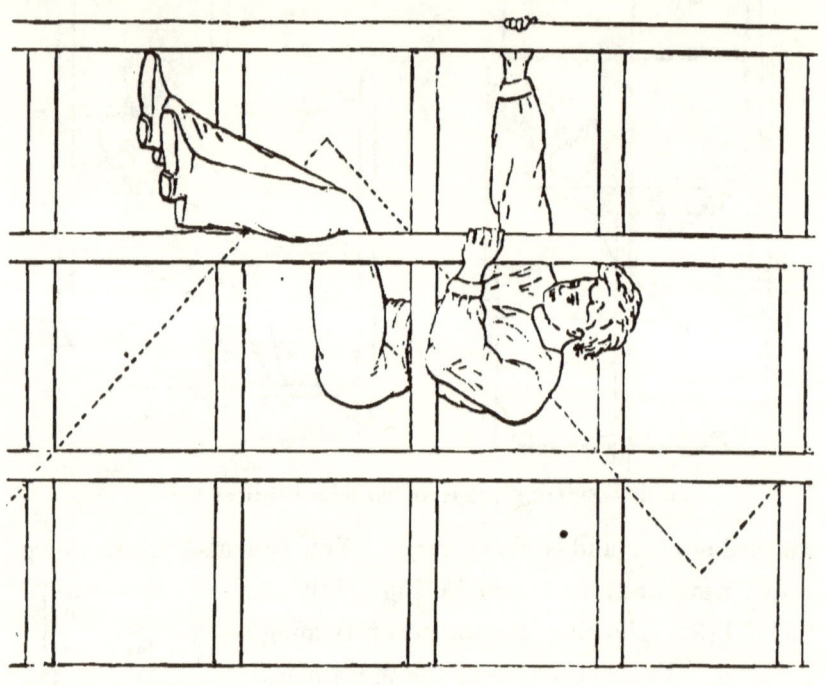

FIG. 121.—ZIGZAG SERPENTINE, GOING DOWNWARD, HEAD FIRST.

head first, whereas in the vertical ladder this takes more time to learn.

An excellent exercise to cultivate the habit of using the opposite limbs simultaneously we have in the ordinary ladder climbing, as shown in Fig. 124. The command is, "*Upward march — óne! Twó!*" ... At the start, both feet are on the floor, and both hands grasp as high a round as they can reach (on the rope-ladder the hands grasp the side-ropes); at "*one*," the left foot is moved one step up, and, at the same time, the right hand grasps above the left one; at "*two*," the right foot moves one

FIG. 122.—*a*, HORIZONTAL SERPENTINE THROUGH LADDER.
b, CORRESPONDING EXERCISE ON DOUBLE BAR.

step above the left, while the left hand grasps above the right one, etc. This movement, which may precede all climbing on

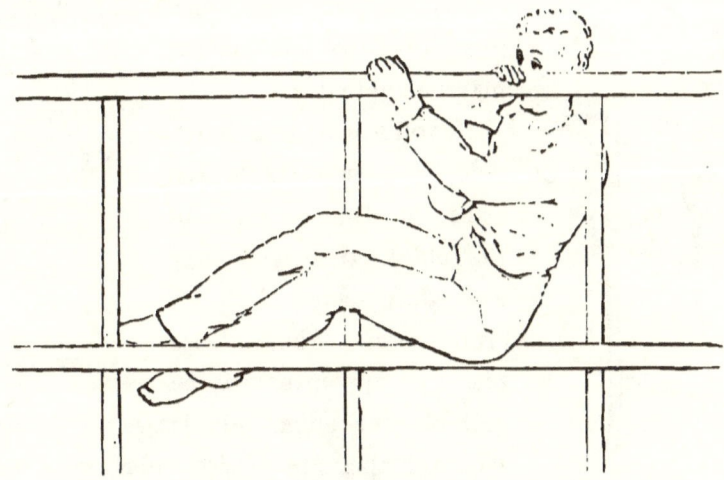

FIG. 123.— HORIZONTAL SERPENTINE BACKWARD, FEET FIRST.

ropes or poles, is one of the easiest of the heaving-movements. It can be done on ordinary vertical (or inclined) ladders, on rope ladders (Fig. 124), and on stall-bars with or without the help of a rope (Figs. 125 and 126).

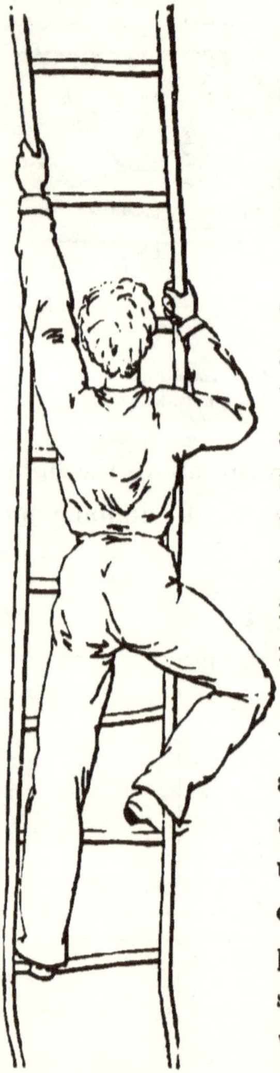

For children, and in absence of other apparatus than stall-bars, some simple substitutes for heaving-movements can be arranged, as shown in Figs. 127, 128, and 129, which hardly need explanation.

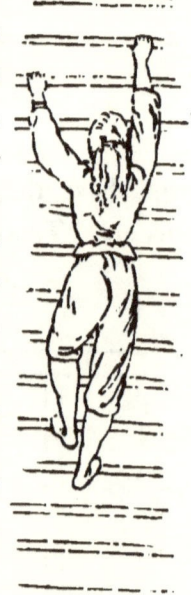

FIG. 125.—CLIMBING THE STALL-BARS.

There is no class of exercises more needed by the growing generations than the heaving-movements, and in any room (schoolroom) some simple contrivance can easily be put up so that at least some form of these movements may be applied.

If you are willing to develop the muscles of the arms at the expense of the chest, then you may use dumb-bells, Indian clubs, etc. These loose pieces of apparatus may be suitable for advanced classes of adults, whose shapes are settled once for all, and who wish to develop muscle and reduce fat, whereas in gymnastics for children they are best left alone, since they have the bad quality of producing and encouraging faulty posture.

FIG. 124.—CLIMB. ROPE LADDER.

MOVEMENTS OF THE SWEDISH SYSTEM 119

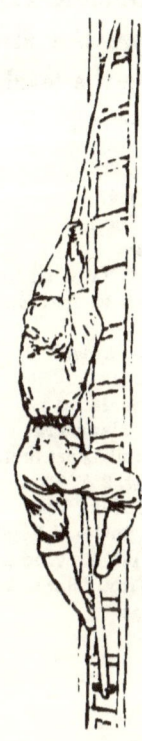

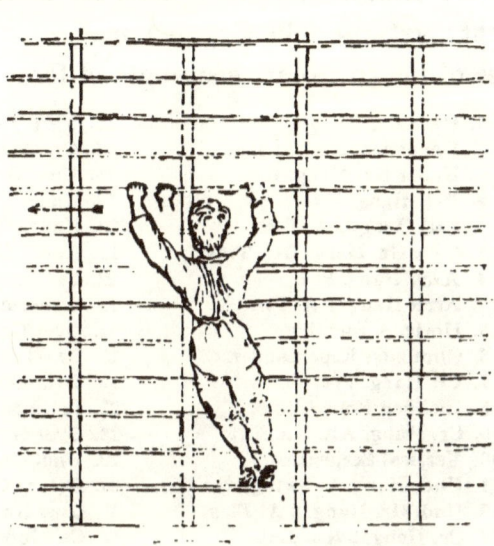

Fig. 127.— Sidew. Trav. on Stall-Bars.

Fig. 126.— Climbing Ladder with Help of a Rope.

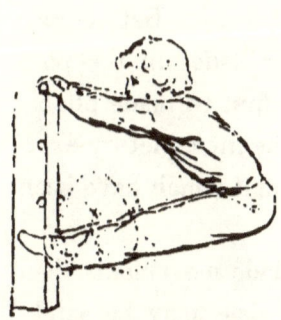

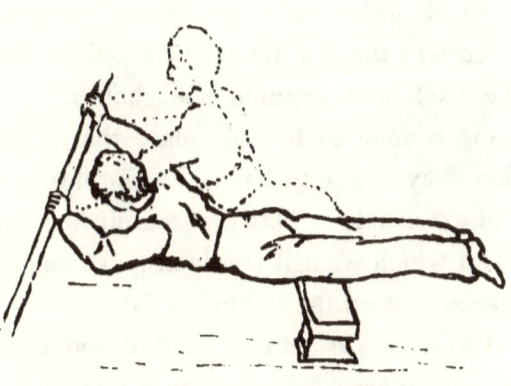

Fig. 128.— Reach Gr. Courtesy Sitt. 2 Kn. Ext.

Fig. 129.— Reach Gr. Forw. Lying 2 A. Ext.

The heaving-movements being of so many different kinds, it is very difficult to include them all in one *progressive* arrangement; yet the following list may serve to give the teacher an idea of the comparative strength of these movements.

1. Over Gr. Hang. ⎫
   Under Hang.   ⎬ Pos.
   Under Gr. Hang. ⎭
2. Fall Hang. Pos.
   Fall Hang. 2 A. Flex.
3. Over Gr. Hang. Osc. Trav.
4. Arch Hang. Pos.
   Arch Hang. 2 A. Flex.
5. Horiz. Serpentine.
6. Climb. on Rope Ladder.
7. Cr. Hang. Pos.
8. Inclined Rope Climb.
9. Cr. Hang. Alt. Kn. Ext.
10. Vertical Serpentine.
11. Und. Hang. L. Swing over Bar.
12. Und. Gr. Hang. 2 A. Flex.
13. Cr. Hang. 2 Kn. Ext.
14. Und. Hang. 2 A. Flex.
15. Over Gr. Hang. 2 A. Flex.
16. Vert. Rope Climb.
17. Cr. Hang. 2 L. Elev.
18. Und. Gr. Hang. Trav.
19. Under Hang. Trav.
20. Under Hang. Vert. Trav. on Ropes.
21. Under Hang. Trav. w. Alt. Elev.
22. Cr. Hang. L. Abd.
23. Somersault betw. Ropes.
24. Over Gr. Hang. Trav.
25. Cr. Hang. Somersault.
26. Over Gr. Hang. 2 Hand Trav.
27. Inclined Rope Climb. Feet first.
28. Over Gr. Cr. Hang. 2 A. Flex.
29. Und. Gr. Hang. 2 Hand Trav.
30. Rotary Climb. on Inclined Rope.
31. Over Gr. Hang. Vert. Trav. on Ladder.
32. Cr. Und. Hang. Trav. (Bar or Ropes).
33. Rotary Trav. on Bar.

SHOULDER-BLADE MOVEMENTS.

By shoulder-blade movements, we mean exercises which tend to correct the posture of the shoulder-blades, flattening them on the back and pushing the shoulders backward. Before anything is done to forcibly place the shoulder-blades in the position they ought to have, the muscles must first be made pliable and extensible. This is accomplished by the introductory exercises, which we call single arm-extensions, and which have been placed first on the following list.

Children are more in need of shoulder-blade movements than grown persons, and two exercises of this class may be safely put into each lesson for children, if the teacher deems such a proceeding necessary.

In the description below, the movements have been grouped

Half Str. St. Pos. (Fig. 130). — Command, "*Left (r.) arm upward stretch — óne! Twó!*" 1. The left (r.) arm takes bend pos. 2. The arm is stretched upward on the same principles as when both arms are stretched (in 2 A. ext.). Common faults are: leaning the head against the arm that is extended upward, and drooping the other shoulder. "*Change arms — óne! Twó!*" 1. Both arms are bent upward (Fig. 80, p. 88). 2. The left (r.) arm is extended downward, the right (l.) one upward.

For children, the movement may be done from bend st. pos. Command. "*Arms upward — bend! Alternate arm-extension upward, beginning by the left (r.) — óne! Twó!*" . . . One arm bends, while the other one extends. In this manner, an alternate arm-extension may also be done sideways or forward.

Half Str. Half Yd. St. Pos. (Fig. 131). — Command, "*Left (r.) arm upward, right (l.) arm sideways stretch — óne! Twó!*" 1. Bend st. pos. 2. The arms extend as directed. Common faults are: raising the arm, which is in yard pos., above horizontal, and carrying it forward; leaning the head against the arm which is extended upward. etc.

Half Str. Half Reach St. Pos. — Command, "*Left (r.) arm*

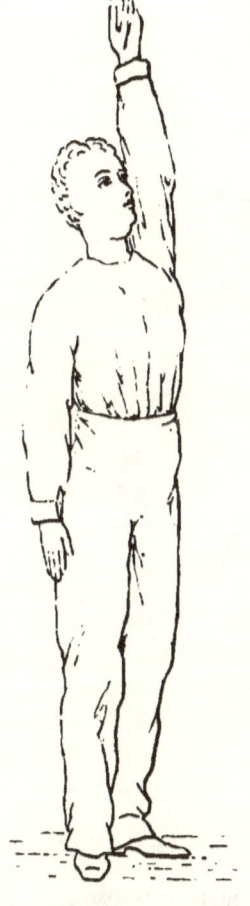

FIG. 130. — HALF STR. ST. POS.

upward, right (l.) arm forward stretch — óne! Twó!" The most common faults in this movement are: stretching the right (l.) arm sideways forward, instead of forward, and throwing the same shoulder forward.

Half Yd. Half Reach St. Pos. (Fig. 132). — Command, "*Left (r.) arm sideways, right (l.) arm forward stretch — óne! Twó!*" Common faults are: twisting the trunk, throwing the shoulders out of position, and not keeping the arms horizontal.

In the last three exercises, the arms change position at the command, "*Change arms — óne! Twó!*"

1. Both arms bend.
2. The arms are stretched in opposite directions. Any of these exercises, once learned, may be suitably used as introductions to a lesson for even an advanced class.

FIG. 131. — HALF. STR. HALF YD. ST. POS.

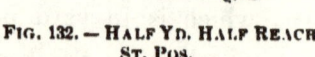

FIG. 132. — HALF YD. HALF REACH ST. POS.

The following movements are shoulder-blade movements of a more specific nature.

St. 2 A. Ext. Backw. (Fig. 133). — Command, "*Arms backward stretch — óne! Twó!*" 1. Bend st. pos. 2. The forearms are flung downward and backward as far as good posture will allow, and stay there, straight, parallel, and with the palms of the hands turned toward each other. Common faults are: pushing the head forward and concaving the chest. Care should be taken that this exercise be not used too early; for, unless performed in very good posture, it has an effect very different from the one desired — which is to expand the clavicular (upper sternal) region of the chest, thus pushing the shoulder-blades backward. To counteract ill results as well as to increase the good ones, the movement is best combined with (or immediately followed by) 2 A. ext. sidew. Command. "*Arms sideways and backward stretch — óne! . . . Fóur! The same — óne! . . . Fóur!*" . . . This combination is much more powerful than either of the two movements alone, as any one will know who has tried it.

FIG. 133. — ST. 2 A. EXT. BACKW.

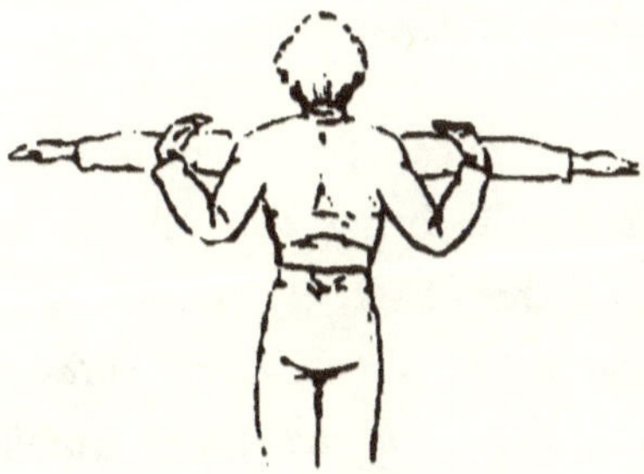

St. Slow 2 A. Ext. to Yard St. *d* **Pos.** (Fig. 134).[1] — Command, "*With palms turned up, arm extension sideways, slowly — one! Two!*" . . . 1. Bend st.

FIG. 134. — ST. 2 A. EXT. TO YARD ST. *d* POS.

[1] Yard st. positions are: (*a*) arms forward bend (Fig. 135); (*b*) arms horizontal, elbows at right angles, upper arms pointing sideways, forearms forward, and hands turned with palms toward each other; (*c*) arms extended sideways, palms turned down; (*d*) arms extended sideways, palms turned up (Fig. 141); (*e*) upper arms horizontal, elbows at right angles, forearms vertical, and palms turned toward each other (Fig. 137).

pos. 2. The arms are extended slowly sideways, with the palms of the hands turned upward. Both the flexion and the extension should be done as evenly as they are slowly. (Not suitable for children under fourteen years of age.)

Yard *a* **St. 2 A. Fling.** (Fig. 135). — Command, "*Arms forward — bénd!*" The arms, bent at the elbows, are lifted sideways to horizontal position with the elbows well drawn backward, the palms of the hands turned down, the forearm and hand in a straight line, and a little below the level of the shoulder. This is the yard st. *a* pos. (Fig. 135 *a*), which should be practised alone before any arm flinging is done. The hands

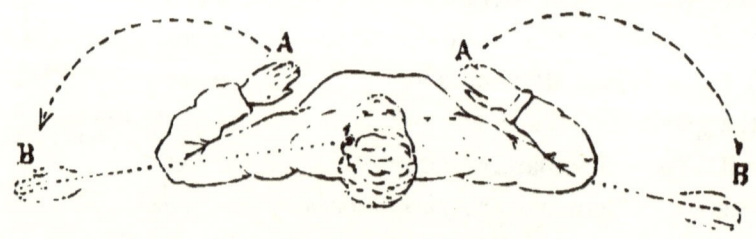

FIG. 135. — YARD *a* ST. 2 A. FLING.

must not touch each other or the chest (as described by a recent German author), as that would bring the elbows forward, and lessen the expansion of the chest. The movement, done from this position, is, "*Arm flinging* (or *Arms sideways fling*) — óne! Two!*" . . . 1. The forearms are flung quickly sideways to yard st. *c* pos. (arms horizontal, palms turned down) (Fig. 135 *b*). 2. With moderate speed (so that they may not strike the chest), the arms resume yard st. *a* pos.

Reach St. 2 A. Fling. (Fig. 136). Command, "*Arms forward — strétch!* (see heaving-movements). *Arm flinging* (or *Arms upward fling*) — óne! Two!*" . . . 1. Remaining straight, the arms are flung quickly upward into stretch st. pos. 2. With

MOVEMENTS OF THE SWEDISH SYSTEM 125

moderate speed, the arms are lowered into reach pos. A common fault is to push the head forward, when the arms rise. To counteract this, beginners may do the movement with the head bent backward. Command, "*Arms forward — strétch! Head backward — bend! Arm flinging — óne! Two!*" . . .

Beginners find it difficult to return to the reach st. pos. without lowering the arms below horizontal; for this reason, the

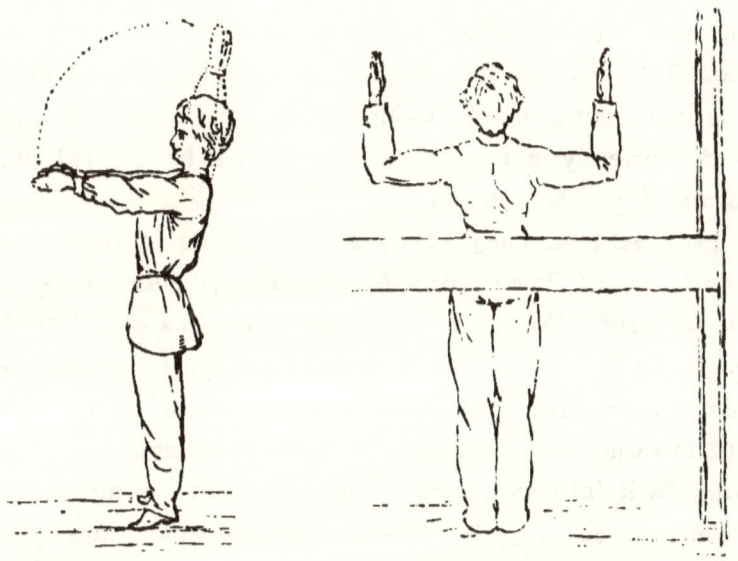

FIG. 136. — REACH ST. 2 A. FLING.   FIG. 137. — YARD *e* SUPPORT ST. POS.

arm flinging may at first be done from fundamental position (st. 2 A. fling. upw.). For children it may be still more simplified by flinging each arm alternately from fundamental position, one arm being flung up, while the other moves down.

Arm flinging upward can also be done from yard *b* st. pos.

Yard *b* St. 2 A. Fling. [to Yd. *e* St. Pos.] (Fig. 137). — Command, "*Arms half forward — bénd!*" This is yard *b* st. pos., in which the arms are horizontal, elbows bent at right angles, and on a level with the shoulders, palms turned toward each other, forearms parallel and pointing forward. The movement to be done from this position is, "*Arm flinging — óne! Twó!*" ... 1. The upper arms remain horizontal and immovable, while the forearms are flung upward until they are vertical (or carried still more backward, if possible). 2. With moderate speed, the forearms are lowered into yard *b* st. pos. When this movement is done for the first time, the pupils may be placed with the small of their backs against a horizontal bar (as shown in Fig. 137). In this way, better muscular isolation is attained. The movement, simple though it be, is exceedingly difficult to do correctly, and consequently it can be applied only to advanced pupils, — never to children.

Yard St. 2 A. Fling. Forward. — Command, "*Arms sideways stretch — one! Two! Arm flinging forward with turning of the hands — óne! Twó!*" ... 1. The forearms are slightly lifted, and while the hands are turned, palms up, the arms are flung sideways forward into reach position (with palms up). 2. The forearms are lifted, and the hands turn, while the arms are flung back into yard pos. The movement resembles cutting the air with two swords. The most common faults are: pushing the head forward, when the arms are flung backward, and bending the arms so much that the movement resembles an arm-extension, the line of motion being a straight line instead of a curve.

Wing Stoop Stride St. Pos. — (For command, etc., see arch-flexions.) In this position the extensors of the back are forcibly contracted, and thereby the shoulder-blades are drawn downward and flattened. To keep the pupils busy, while standing in this position, a head-rotation may be ordered.

Stoop Stride St. H. Rot. (Fig. 138). — After the position is taken, command, "*Head rotation, to the left (r.), right (l.) and forward — óne! Twó! Thrée! The same right (l.), left (r.) and forward — óne! ... Threé!*" etc.

Other exercises in this position are

Bend Stoop Stride St. 2 A. Ext. Upw. (Fig. 139), at first done quickly, later on very slowly. When done for the first time, the arms should be stretched upward before the trunk is bent forward, as the str. stoop stride st. pos. (Fig. 59, p. 73) makes it easier to extend the arms (not only upward, but also in other directions). For beginners the arm-extension may be done alternately from bend. pos., as described above. We also have

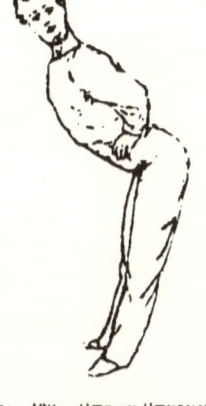

FIG. 138.—STOOP STRIDE ST. H. ROT.

Stoop Stride St. 2 A. Ext. Upw. and Sidew.;
Half Yd. Hlf. Str. Stoop Stride St., Change of A.;
Reach Stoop Stride St. 2 A. Fling. (usually beginning from str. pos.);
Yard *a* Stoop Stride St. 2 A. Fling.;
Bend Stoop Stride St. Slow 2 A. Ext. to Yd. *d* Pos.; the commands and execution of which movements we hardly need to describe.

Yard Stoop Stride St. 2 A. Circ. — Command, "*Feet sideways place and arms sideways stretch — óne! Twó! Trunk forward — bend!*" [This trunk-flexion alone is a suitable shoulder-blade movement for children (Fig. 140).] "*Arm circumduction — start!*" The arms remain straight, and are moved with moderate speed so that the hands describe a small circle upward, backward,

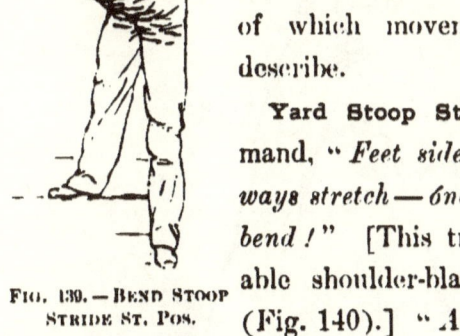

FIG. 139.—BEND STOOP STRIDE ST. POS.

128   *EDUCATIONAL GYMNASTICS*

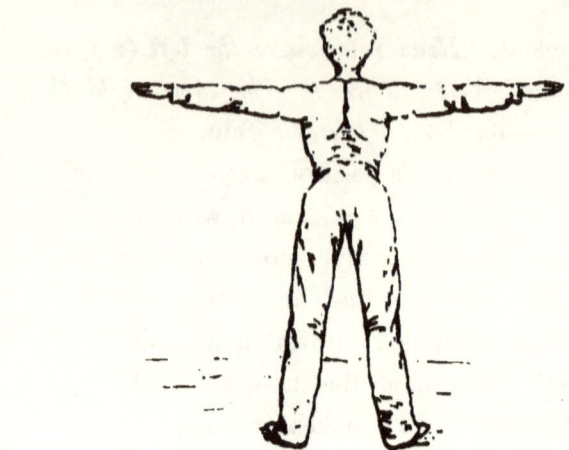

FIG. 140. — YARD STOOP STRIDE ST. POS.

FIG. 141. — YARD *d* STOOP STRIDE ST. 2 A. ELEV.

downward, and forward, and this is continued until the teacher commands "*Stóp!*"

Yard *d* Stoop Stride St. 2 A. Elev. (Fig. 141). — Command, "*Left (r.) foot sideways place and arms sideways — strétch! Trunk forward — bend! Hands — túrn! Arm elevation — one! Two!*" . . . 1. The arms are slowly raised sideways into str. pos. 2. The arms resume yard *d* pos. equally slowly. At first the movement is best done from str. stoop stride st. pos., the command then being (after commencing position is taken), "*Arms sideways sink and upward lift — one! Two!*" . . . In this form, the movement is easier. For advanced pupils it may be changed to

Half Yd. Hlf. Str. Stoop Stride St. A. Elev., in which movement one arm is lifted, while the other is lowered, so that the hands maintain the same distance from each other as in the commencing position.

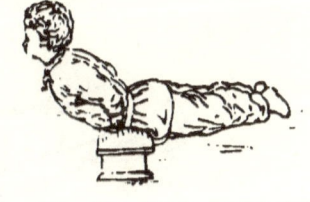

FIG. 142. — WG. FORW. LYING POS.

Any of the above movements can be done as well from stoop st. pos. (with heels together).

Forward Lying Pos. (Fig. 142). — The pupil lies down across a bench so that his waist is above the middle of the bench; his legs are straight, and his feet and hands rest on the floor. At the command "*Hips — fírm!*" he places his hands on his hips, and raises his trunk as high as he can, his feet leaving the floor so that he balances on the bench. At the command "*Position!*" he resumes commencing position. When movements are to be done with this for a commencing position, the legs of the pupil are prevented from rising above horizontal, by another pupil kneeling behind and grasping the first one above and around the ankles (as in Figs. 143 and 144). In a schoolroom this may be arranged so that the pupil lies down on the desk, and places his feet under the cross-tree of the back rest of his chair.

Movements in this position are

Forward Lying H. Rot.;

Rest Forw. Ly. T. Backw. Flex.;

Str. Forw. Ly. T. Backw. Flex. — In the last two, after commencing position is taken, the command is, "*Trunk forward — bend!*" when the pupil bends until he touches the floor; and "*Trunk backward — bend!*" when he rises as high as he can. Other movements are

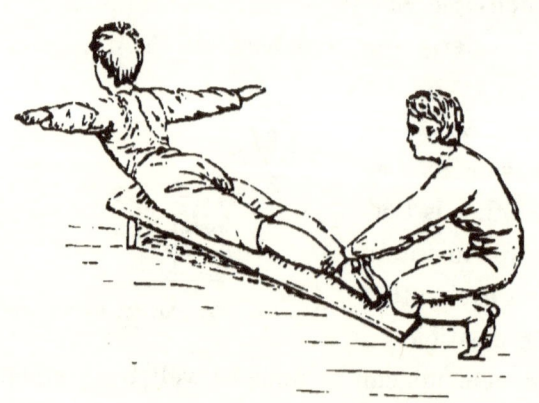

FIG. 143. — YARD c FORW. LY. POS.

Yard Forw. Ly. 2 A. Fling. (Fig. 143);

Forw. Ly. 2 A. Ext. Sidew. (1. quickly, 2. slowly);

Forw. Ly. 2 A. Ext. Upw. (Fig. 144);

Yard *d* **Forw. Ly. 2 A. Elev.** These will need no description.

All of these, except Nos. 3, 6, and 7, are especially suitable for children. The farther forward the pupil is lying, the stronger is the effect of the movement. The most common fault is pushing the head forward.

Half Str. Fallout St. *a* **Change of A.** — Command, "*Left (r.) arm upward, right (l.) arm backward, and to the left (r.) fallout — óne! Twó!*" 1. Bend st. pos. 2. The left (r.) arm is stretched upward, the right (l.) one backward, while the fallout st. *a* pos. is taken (as described under leg-movements) (Fig. 145).

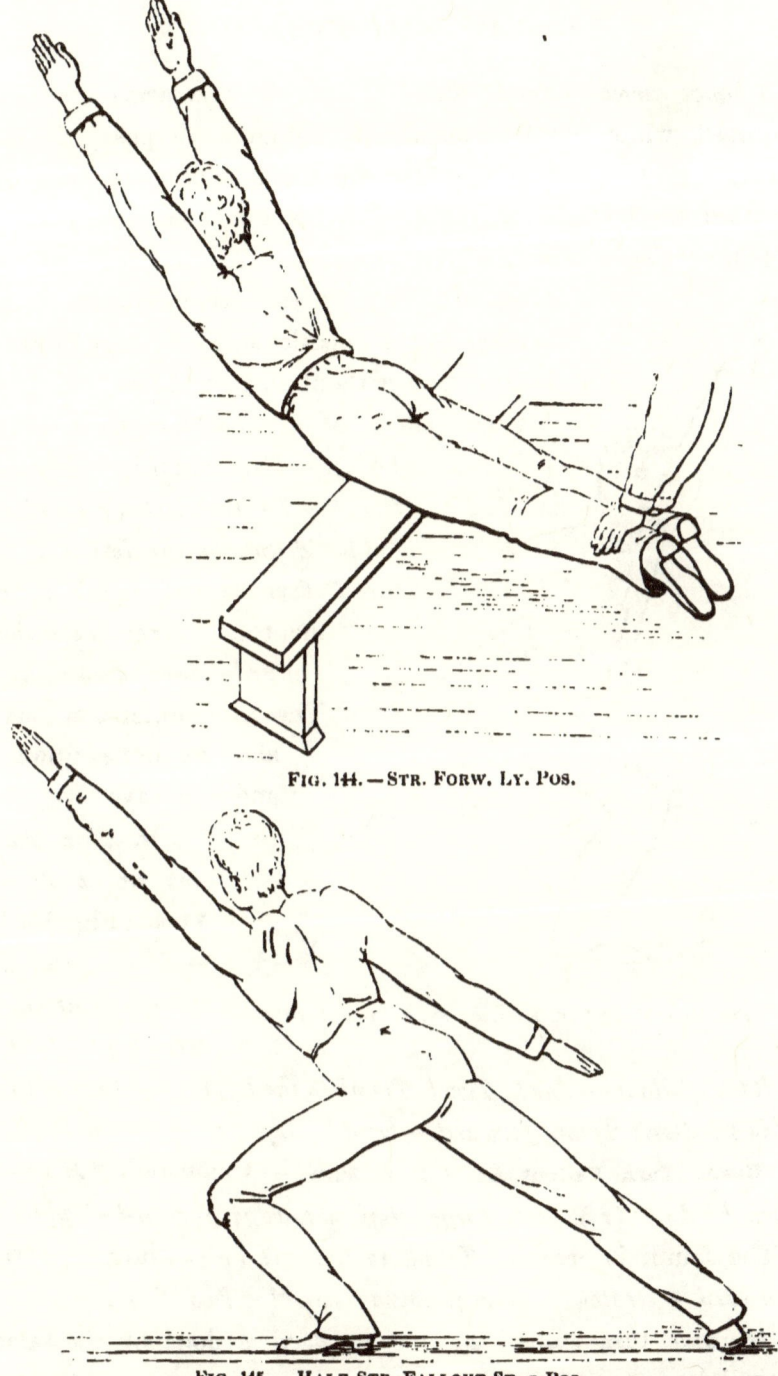

Fig. 144. — Str. Forw. Ly. Pos.

Fig. 145. — Half Str. Fallout St. *a* Pos.

"*Change arms — óne! Twó!*" ... 1. The arms are bent upward, while the feet remain in fallout st. *a* pos. 2. The arms are stretched in opposite directions, while the trunk is rotated to the side of the arm that extends backward (Fig. 146). "*Change arms and feet — óne! Twó!*" 1. Bend. st. pos. (Fig. 80, p. 88). 2. Hlf. str. fallout st. *a* pos. is taken to the opposite side. Change of arms is now done without change of feet, etc.

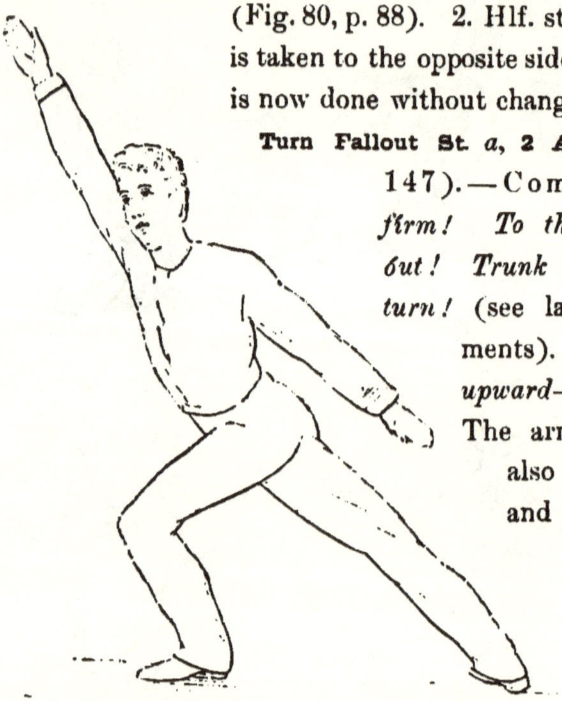

Turn Fallout St. *a*, **2 A. Ext. Upw.** (Fig. 147). — Command, "*Hips — firm! To the left (r.) — fall-óut! Trunk to the left (r.) — turn!* (see lateral trunk-movements). *Arm extension upward — óne! Twó!*" ... The arm extension may also be done upward and sideways.

Str. Turn Fallout St. *a* **Forw. Flex.** (Fig. 148). — Command, "*Arms upward stretch and to the left (r.) fallout — óne! Two! Trunk to the left (r.) — turn!* (In this position) *Trunk forward — bend! Upward — stretch!*" ...

Reach Turn Fallout St. *a* **2 A. Fling.** — Command, "*Hips — firm! Left (r.) foot, large step, sideways forward — pláce!* (The trunk is erect.) *Trunk to the left (r.) — turn! Arms forward — stretch! Arm flinging — óne! Two!*" ...

Yard *d* **Turn Fallout St.** *a* **2 A. Elev.** is done on the same principles.

Fig. 146. — Half Str. Fallout St. *a*, Change of A.

MOVEMENTS OF THE SWEDISH SYSTEM 133

FIG. 147.—STR. TURN FALLOUT ST. *a* POS.

FIG. 148.—STR. TURN FALLOUT ST. *a* FORW. FLEX.

In a similar manner, though without rotation of the trunk, are executed:

Fallout st. *b*.	2 A. ext. upw. (sidew. & upw.) (Fig. 150).
Fallout st. *c*.	2 A. fling.
Toe sup. fallout st. *b*.	2 A. elev. (Fig. 149).
Foot gr. fallout st. *b*.	T. forw. flex. (Fig. 151).

Besides these we also have:

Str. Fallout St. *c* **2 A. Ext. w. Change of Feet Backward.** — Command, "*Arms upward stretch, feet close and left (r.) foot*

FIG. 149. — TOE SUP. FALLOUT ST. *b* 2 A. EXT. UPW.

FIG. 150. — BEND FOOT GR. FALLOUT ST. *b* POS.

backward — fallóut!" (The feet are brought together, when the arms bend.) "*Arm-extension upward with change of feet — óne! Twó!*" . . .

Str. Foot Gr. Fallout St. *b* **Kn. Flex.** — After the pupils are arranged at the stall-bars, command, "*Hips — fírm! Left (r.) foot on the third bar — place! Forward — fallóut!*" (see Fig. 26). *Arms upward — strétch! Knee — bend!*" The knee which is bent bends still more, while the other (backward) one remains straight. *"Stretch!"* The forward knee is straightened until it has its original flexion (about 90°). **Str. Hor. Half St. Kn. Flex.** (Fig. 152). — Command. *"With arm-extension upward, horizontal half-standing position on the left (r.) foot place — óne! Two!"* 1. Bend st. pos. 2. The arms are extended upward, and, at the same time, the left (r.) knee bends, the body inclines forward, and the right (l.) leg is lifted backward, so that the arms, trunk, and backward leg form an even, nearly horizontal curve. *"Knee — bend!"*

FIG. 151. — STR. FOOT GR. FALLOUT ST. *b* FORW. FLEX.

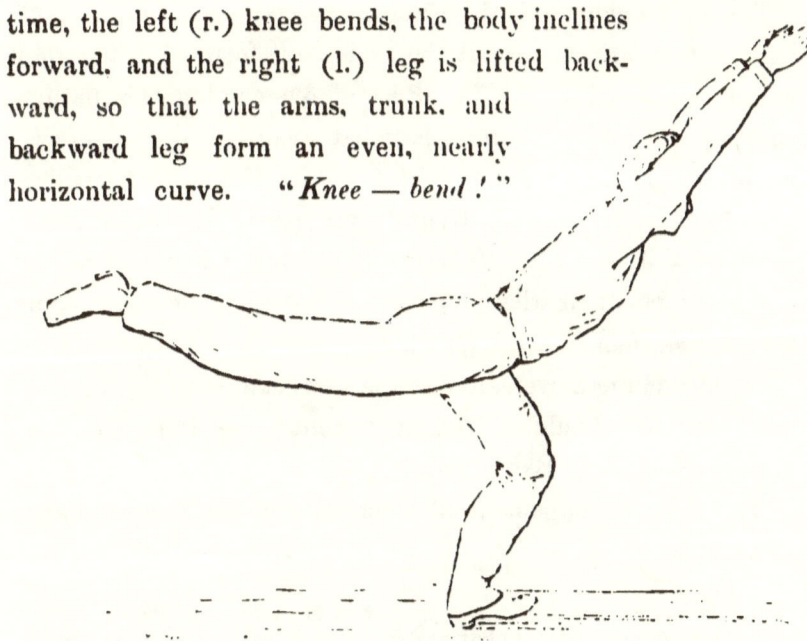

FIG. 152. — STR. HOR. HALF ST. POS.

The knee of the supporting leg bends to utmost flexion, the rest of the body maintaining the same position. *"Stretch!"*

The knee resumes original flexion. At first this movement (which is prepared by the corresponding leg-movements; see Fig. 49, p. 64, etc.) is practised in half str. pos., the other hand taking "hip — firm," either so that the left arm is extended when the left leg is raised, or so that the elevated arm and leg belong to opposite sides. The knee-flexion is not tried until the commencing position can be correctly taken. Very advanced pupils may try an arm-extension (2 A. ext. upw.) in this position: **Hor. half st. 2 A. ext. upw.** This movement and the one just described are exceedingly difficult on account of their close relationship to the balance-movements — in which class they might also be counted.

The most common fault in these exercises in fallout st. positions, or their allies, is to let the arms sink forward, when they should be in str. pos. The effort of overcoming this natural tendency to faulty posture is what makes these exercises shoulder-blade movements.

If dumb-bells or other weights are used, the safest movements are 2 A. elev. and 2 A. ext. in stoop stride st. and fallout st. *a* and *b* positions, the danger of compressing the chest being less in these than in other positions.

The swimming-movements of the arms can, with slight variations, serve as shoulder-blade movements (see Part III., Preparatory Swimming Exercises).

Progressive arrangement of a majority of the shoulder-blade movements:

1. Hlf. Str. St. change of arms.
2. Hlf. Str. St. A. Fling. (small children).
3. Bend St. Alt. A. Ext. Upw. ⎫ children
4. Yard St. *d* Alt. A. Flex. ⎬ of gram-
5. Bend St. Alt. A. Ext. Forw. ⎭ mar age.
6. Hlf. Str. Hlf. Yd. St., change of arms.
7. Wing Stoop Stride St. H. Rot.
8. 2 A. Ext. Sidew. and Backw.
9. St. 2 A. Fling. Upw. (prim. and gram.).
10. Yard Stoop Stride St. H. Rot.
11. Hlf. Str. Hlf. Reach St. Pos.
12. Yard Stoop Stride St. 2 A. Circ. (children 14–17).
13. Hlf. Yard Hlf. Reach St. Pos.

MOVEMENTS OF THE SWEDISH SYSTEM 137

14. Yard Stoop Stride St. 2 A. Fling.
15. Hlf. Str. Fallout St. *a*, change of arms.
16. Str. Stride St. T. Forw. Flex.
17. Yard St. 2 A. Fling. Forw.
18. Reach St. 2 A. Fling.
19. Bend Stoop Stride St. Alt. A. Ext. (children 14-17).
20. Str. Stoop Stride St. 2 A. Ext. Upw.
21. Forw. Ly. Pos.
22. Hlf. Str. Hlf. Yd. Stoop Stride St., change of arms.
23. Forw. Ly. H. Rot.
24. Reach Stoop Stride St. 2 A. Fling.
25. Bend Forw. Ly. 2 A. Ext. Sidew.
26. Str. Fallout St. *b* 2 A. Ext. Upw.
27. Yard Forw. Ly. 2 A. Fling.
28. Rest Forw. Ly. T. Flex.
29. Bend Toe Sup. Fallout St. *b* 2 A. Ext.
30. Str. Turn Fallout St. *a* 2 A. Ext.
31. Reach Fallout St. *b* 2 A. Fling.
32. Bend Stoop Stride St. 2 A. Ext. to Yd. *d*.
33. Str. Stoop Stride St. 2 A. Ext. to Yd. *e*.
34. Yd. Stoop Stride St. 2 A. Elev.
35. Bend Foot Gr. Fallout St. *b* 2 A. Ext.
36. Bend Forw. Ly. 2 A. Ext. to Yd. *d*.
37. Yard Turn Fallout St. *a* 2 A. Elev.
38. Str. Fallout St. *b* Forw. Flex.
39. Reach Turn Fallout St. *a* 2 A. Fling.
40. Str. Foot Gr. Fallout St. *b* Kn. Flex.
41. Rest Foot Gr. Fallout St. *b* Kn. Flex.
42. Str. Foot Gr. Fallout St. *b* 2 A. Fling. Upw.
43. Str. Foot Gr. Fallout St. *b* Forw. Flex.
44. Str. Turn Fallout St. *a* Forw. Flex.
45. Yd. *b* 2 A. Fling. to Yd. *e*.
46. Rest Foot Gr. Fallout St. *b* Forw. Flex.
47. Bend Forw. Ly. Alt. A. Ext. Upw. (children).
48. Bend Forw. Ly. 2 A. Ext. Upw.
49. Str. Fallout St. *c* 2 A. Ext. Upw.
50. Str. Hor. Half St. Kn. Flex.
51. Yd. *d* Foot Gr. Fallout St. *b* 2 A. Elev.
52. Str. Forw. Ly. Forw. Flex.
53. Yd. *d* Fallout St. *c* 2 A. Elev.
54. Str. Fallout St. *c* Forw. Flex.
55. Str. Fallout St. *c* 2 A. Ext. w. change of feet backw.
56. Yard *d* Forw. Ly. 2 A. Elev.

ABDOMINAL EXERCISES.

Exercises that chiefly bring into play the muscles of the abdomen, we call abdominal exercises. These movements are nearly related to the arch-flexions, and, in some forms, the two classes merge into each other, so that the distinction between them can hardly be made. The effect of these exercises is to strengthen the abdominal walls, and thus, secondarily, to affect the digestive organs. Children will need these movements less than grown persons, and whereas the abdominal exercises may be entirely omitted from the lessons for children, two might be applied in one lesson for grown persons. Having a tendency to contract the chest, these movements should not be introduced until arch-flexions and heaving-movements have been practised a sufficiently long time (six or eight weeks) to insure a tolerably good posture of the chest. The abdominal exercises are comparatively few and, at the same time, very powerful, so that the progression must be slow and the same exercise may be conveniently repeated through two or more consecutive tables.

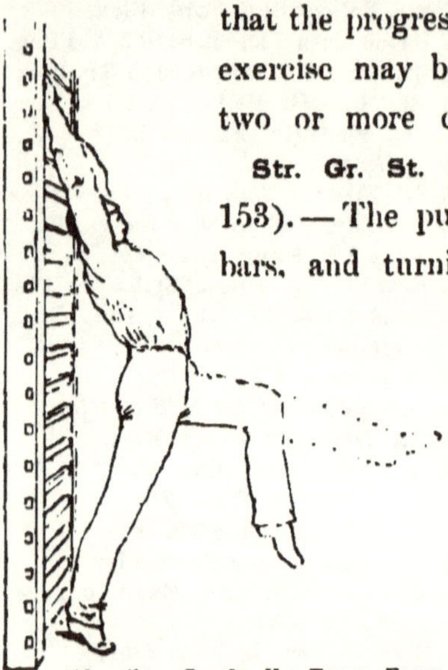

Str. Gr. St. Kn. Forw. Flex. and Ext. (Fig. 153). — The pupil stands close up to the stall-bars, and turning his back to them. At the command, "*Arms upward — strétch! Grásp!*" he stretches his arms up, grasps a bar as high as he can reach, and lets his body arch forward so that arms and legs are straight. Command, "*Left (r.) knee upward — bend! Knee forward — stretch! Bend! Stretch!*" . . . *Change feet — óne! Two!*" . . . just as for the corresponding free-standing exercise (see leg-movements), which it resembles in execution. A doorway, the vertical poles, etc., are just as useful as apparatus for this movement.

Fig. 153. — Str. Gr. St. Kn. Forw. Flex. and Ext.

Str. Lying 2 L. Elev. (Fig. 154). — The pupil lies down flat on the floor (face up), and after the teacher has ordered stretch position, he commands, "*Leg elevation — one! Two!*" . . . (or "*Legs — lift! Sink!*" . . .)

Fig. 154. — Str. Lying 2 L. Elev.

With straight knees and extended insteps, the pupil raises his legs slowly and lowers them as slowly. At first the elevation is very slight — just enough to be called an elevation —

a few days later it is done to 90° (or over), and still later it is done to 45°, where a distinct pause is made before the legs are again lowered. In this way, this movement has a progression in itself. The amount of elevation is indicated by adding "*slightest elevation,*" "*all the way up,*" or "*half way,*" to the preparatory part of the command. The tendency to tip the arms up can be overcome if the hands grasp the horizontal bar put low (or other apparatus, Fig. 155), or by having the hands held down by another pupil. If the floor is unsuitable and benches are at hand, the pupils lie down on these instead, lengthwise, one pupil at each end, their hands grasping the edges of the bench (Fig. 156).

Fig. 155.

Kn. St. Pos. (Fig. 157).— To take the knee standing position, first get into courtesy sitting position (leg-movements, Fig. 39, p. 58), and at the command, "*Kneel!*" move the knees forward until they rest on the floor; the balls of the feet should still rest on the floor, so that standing position can be resumed by merely moving the centre of gravity backward, and then extending the legs. The trunk should be erect.

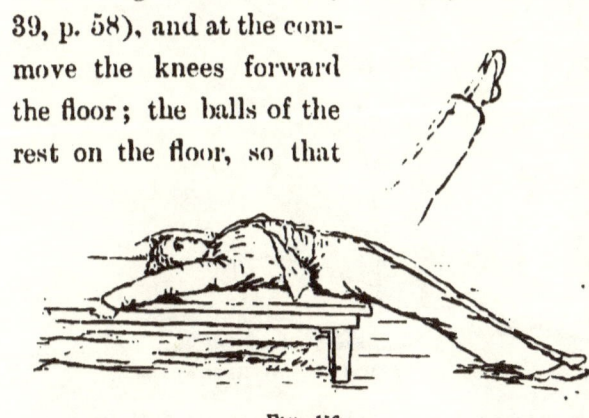

Fig. 156.

Half Kn. St. Pos. (Fig. 158).— To take this position, one leg is moved backward as in toe support fallout st. *b* pos., and the knee is bent until it rests on the floor. Meanwhile the other knee is bent so that the forward foot rests fully on the floor; the ball of the backward foot is also resting there,[1] and the trunk is erect. The command is, "*Half knee standing position, on the left (r.) knee — pláce!* Change *knees — pláce!* (or — *óne! Twó!*)." These two are commencing

FIG. 157. — STR. KN. ST. BACKW. FLEX.

positions for various abdominal exercises enumerated below. Movements in the half Kn. st. pos. are slightly stronger than the same ones in the Kn. stand. position. The movements in these positions are as follows:

FIG. 158. — STR. HALF KN. ST. BACKW. FLEX.

[1] Some instructors claim that the backward foot should be resting with the toes on the floor, the instep being stretched. This form may be more æsthetic, but affords less balance and less ability to resume standing position.

MOVEMENTS OF THE SWEDISH SYSTEM   141

Wing-Bend-Rest-Str.- { Kn. st. or Hlf. Kn. st. } T. backw. flex. (Figs. 157 and 158).

Bend arch { Kn. st. Hlf. Kn. st. } 2 A. ext. upw.

Reach arch { Kn. st. Hlf. Kn. st. } 2 A. fling. (Fig. 159).

And they will hardly need description, since the same movements have been described with other commencing positions (arch-flexions, shoulder-blade movements).

Foot. Gr. Sitt. Pos. (Fig. 160, etc.). — The pupils sit on the benches facing the stall-bars, and place their feet between the second and third bars; the trunk is erect and the chest expanded. The benches are at such a distance that the pupils have straight knees. If tall pupils happen to sit beside short ones, the bench should accommodate the short ones, and the tall ones adjust the distance by spreading their feet apart. In absence of stall-bars, the horizontal bar forms a good grasp for the feet (Fig. 161). In total absence of apparatus, the pupil may sit on the floor, another pupil holding his feet (Fig. 162). In a school-room, the pupil may sit on the desk and place his feet under the seat (or back rest) of his chair (which he faces). The movements in this position are:

FIG. 159. — REACH ARCH KN. ST. 2 A. FLING.

Wg.-Bend-Rest-Str.- } Sitt. T. Backw. Flex.

142 EDUCATIONAL GYMNASTICS

Fig. 160.—Str. Foot Gr. Sitt. T. Backw. Flex.

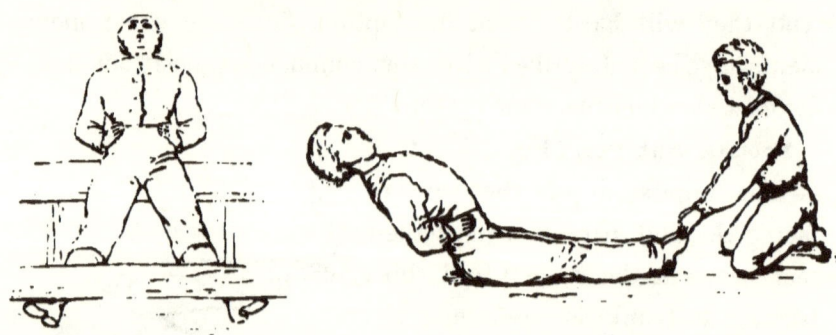

Fig. 161.— Wg. Foot. Gr. Sitt.
T. Backw. Flex.

Fig. 162. — Wg. Foot Gr. Sitt. T. Backw.
Flex.

Fig. 163.— Bend Arch Foot Gr. Half St. 2 A. Ext. Upw.

MOVEMENTS OF THE SWEDISH SYSTEM 143

The flexion has a progression in itself; viz., 1. Slightest flexion; 2. Flexion till the floor is touched; 3. Flexion to horizontal position. (Compare str. lying 2 L. elev.)

Bend Arch Sitt. 2 A. Ext. Upw. and **Reach Arch Sitt. 2 A. Fling.** are executed as described above (see arch-flexions and shoulder-blade movements).

FIG. 164.

Foot Gr. Half St. Pos. (Fig. 165). — The pupil stands facing the stall-bars and one step away from them; he places one foot

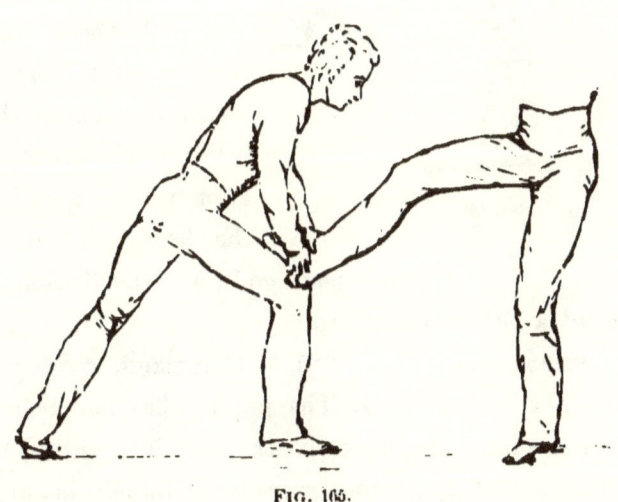

FIG. 165.

forward between the third and fourth bars, and keeps both knees straight; the trunk is erect. In absence of stall-bars, the lifted foot may be placed on a bench (on a chair; on the knee of another pupil [Figs. 164 and 165]), and held there by another pupil. Movements in this position are

Wg.-
Bend- } Foot Gr. Half St. T. Backw. Flex.;
Rest-
Str.-

Bend Arch Foot Gr. Half. St. 2 A. Ext. Upw.;
Reach Arch Foot Gr. Half St. 2 A. Fling.

In all these movements, both knees are kept straight, and the body is bent so far back that it is in a line with the raised leg.

If this leg is raised to horizontal position (or above the fourth bar), the knee of the supporting (backward) leg must bend as the trunk is bent backward (Fig. 166).

FIG. 166.

This form, however, is used only for very advanced classes.

In all these backward flexions, it is very important not to check the respiration. The chest must be kept well expanded, and the head held well back. ("Bréathe!" "Let your head go!") The flexion has the character of a *falling* backward.

Stoop Falling Pos. (Fig. 168). — Command, "*Stoop falling position — óne! Twó!*" 1. The pupil takes courtesy sitting position, and places his hands on the floor in front of him, the hands about a foot apart, and pointing obliquely inward (Fig. 167). 2. Keeping his body straight, and his hands in the same place, the pupil throws his feet backward so that his legs become straight, and in a line with his body. The under side of the toes should rest on the floor (not the upper side, as described by some authors. Compare Kn. st. pos.). To return to standing position, command, "*Position — óne! Twó!*" 1. The feet are placed forward, behind (or between) the hands as above. 2. Fundamental position is resumed.

Fig. 167.

To take the stoop falling pos., is in itself an abdominal exercise, as it could not be maintained were it not for the strong contraction of the abdominal walls. If it is desirable to make the pupils stay for some time in this position, keep them busy by commanding a rapid head-rotation:

Stoop Fall. H. Rot. (Commands, etc., as described.)

To teach the pupils to quickly take the stoop fall pos., the moving of the feet backward and forward may be practised as a separate movement. This also furnishes a short rest for the abdominal muscles when the movement (position) is to be made stronger by longer duration:

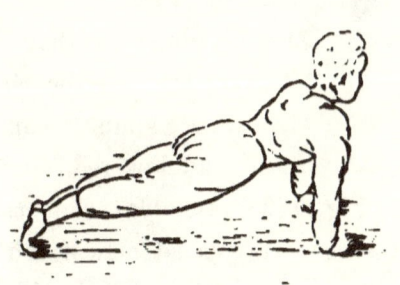

Fig. 168.— Stoop Falling Position.

Stoop Fall. F. Placing Forw. and Backw. — Command, "*Stoop falling position — óne! Twó! Feet forward — pláce! Backward — pláce! The same — óne! Twó!*" . . .

In this position the following movements may be done:

Stoop Fall. 2 A. Flex. — After commencing position is taken, command, "*Arms — bend!*" The body remains rigid, and the arms bend until the nose touches the floor. "*Stretch!*" The arms are straightened.

Stoop Fall L. Elev. — After commencing position, command, "*Leg elevation — óne! ... Fóur!*" 1. The left (r.) leg is lifted with straight knee. 2. The left (r.) leg is lowered, and the foot again put on the floor. 3, 4. The right (l.) leg is raised and lowered. This exercise can also be done while the arms are bent, and is then called

Bend Stoop Fall. L. Elev.

Stoop Fall. A. Elev. — For command and description, see stoop fall L. elev., and exchange the words "leg" for "arm," "knee" for "elbow," and "foot" for "hand."

Stoop Fall. L. and A. Elev. is done so that the left (r.) leg and the right (l.) arm are simultaneously raised and lowered, etc.

FIG. 109. — HORIZONTAL STOOP FALL. POS.

Horizontal Stoop Fall. Pos. (Fig. 169) resembles the stoop falling position, except that the feet are placed on a bench or other apparatus, so that they are on a level with the shoulders. If the feet are placed still higher (on stall-bars, horizontal bar, etc.), the position is called

Inclined Stoop Fall. Pos. (Fig. 170).

In either position, double arm-flexion can be done, and for moderate elevation head-rotation also. The inclined stoop fall.

pos. can be taken from stoop fall. pos. by having another pupil grasp the feet of the one doing the movement and lifting them from the floor (Fig. 171).[1] This form, however, is suitable only for advanced pupils and grown persons.

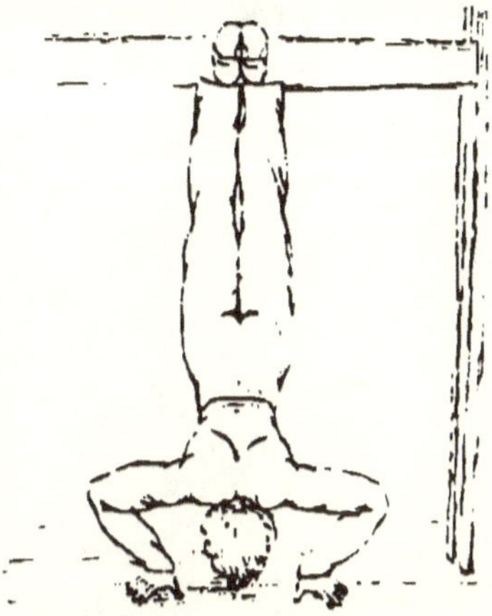

If the legs are raised beyond vertical position and resting forward against a wall (stall-bars, or other apparatus), the position is called

Reverse Stoop Fall. Pos. (Fig. 173). — The pupil stands facing the stall-bars (or wall) and two steps away from them, one foot behind the other (walk st. *b* pos.) (Fig. 172). At the command,

FIG. 170.—INCL. STOOP FALL. 2 A. FLEX.

"*Reverse stoop falling position — place!*" he places both hands on the floor (as in stoop fall. pos.) and swings his legs up until his feet rest against the wall, the weight of the body being carried by the straight arms. When the head is carried well up and the back is arched, there is no danger of a fall. In this position, arm-flexion can be done to the command, "*Arms — bend!*

FIG. 171.

[1] Fig. 171 is faulty, inasmuch as the pupil whose hands are on the floor shows these turned out instead of obliquely inward.

Stretch!" For more safety, the feet may be inserted between the bars. This position, naturally, is to be tried only by far-advanced pupils; those who have a flexible back may move the feet down from bar to bar until the feet stand on the

FIG. 172. — POSITION PREPARATORY TO REVERSE STOOP FALLING.

floor in front of the head (as in gr. arch st. pos. w. hands on the floor, Fig. 76, p. 84), which position might be named **rev. stoop fall. w. feet on the floor**. The rev. stoop fall. pos. may also be done with living support, one pupil standing in front of the other and grasping the feet when they swing up (Fig. 174), or, still better, the legs near the ankles.

The heaving-movements in cr. hang. pos. must be considered

FIG. 173. — REVERSE STOOP FALL. POS.

as strong abdominal exercises as well; and whenever a heaving-movement of this kind is used, there need be no other abdominal exercise in the same lesson.

For children, leg-elevation in fall. hang. pos. is suitable as an abdominal exercise; the arms may be straight or bent (Fig. 175). After fall hang. pos. is taken, command, "*Beginning with the left (r.), leg elevation — óne! . . . Fóur!*" The legs are raised and lowered on the same principles as in stoop fall. L. elev.

In the schoolroom, the stoop fall. pos. may be done by having the pupil stand in the aisle facing the back of the room;

Fig. 174.—Reverse Stoop Fall. Pos. with Living Support.

Fig. 175.— Bend Fall. Hang. L. Elev.

placing a hand on the chair on either side of him, he swings his feet backward at the teacher's command. (For cr. hang. pos. in schoolrooms, see heaving-movements.)

Progressive arrangement of abdominal exercises:

1. Stoop Fall. H. Rot.
2. Str. Gr. St. Kn. Flex. and Ext.
3. Stoop Fall. F. Plac. Forw. and Backw.
4. Str. Ly. 2 L. Elev.
5. Wing Knee St. Backw. Flex.
6. { Wing- / Bend- } Foot Gr. Sitt. Backw. Flex.
7. Hor. Stoop Fall. H. Rot.
[8. Balance Cr. Hang. Pos. (in schoolroom).]
9. Bend Arch Foot Gr. Sitt. 2 A. Ext. Upw.
10. Wing. Foot Gr. Half St. Backw. Flex.
11. Reach Arch Foot Gr. Half St. 2 A. Fling.

12. Rest Foot Gr. Sitt. Backw. Flex.
13. } Rest { Kn. St.   } Backw. Flex.
14. }   { Kn. Half St.   }
15. Stoop Fall 2 A. Flex.
16. Str. Knee St. Backw. Flex.
17. } Bend { Foot Gr. Half St. Backw.
    } Rest { Flex.
18. Str. Knee Half St. Backw. Flex.
19. Hor. Stoop Fall. 2 A. Flex.
20. } Bend Arch { Kn. St.   } 2 A. Ext.
21. }   { Kn. Half St. } Upw.
22. Incl. Stoop Fall. 2 A. Flex.
23. } Reach Arch { Kn. St.   } 2 A.
24. }   { Kn. Half St. } Fling.
25. Bend Arch Foot Gr. Half St. 2 A. Ext.
26. Str. Foot Gr. Half St. Backw. Flex.
27. Reach Arch Foot Gr. Half St. 2 A. Fling.
28. Wing Foot Gr. Half St. Deep Backw. Flex.
29. Stoop Fall. L. Elev.
30. Stoop Fall. 2 A. Flex. and L. Elev.
31. Str. Foot Gr. Sitt. Backw. Flex.
32. Rev. Stoop Fall. Pos.
33. Rev. Stoop Fall. 2 A. Flex.
34. Rev. Stoop Fall. w. hands on the floor.
35. Same position with 2 A. Flex.

LATERAL TRUNK-MOVEMENTS.

These consist of various movements bringing into play the muscles around the waist and those of the lateral halves of the trunk. The effect of these movements is to develop the aforesaid muscles; but that is not all. For the location of the largest blood-vessels in the body is such that these vessels must follow the movements of the spine. If the trunk be bent in any direction and again straightened, the blood-vessels referred to go through a similar motion; and the result of this is an acceleration of the blood contained in them. This secondarily affects all the organs in the neighborhood, and especially those of the digestive system. Hence, although these exercises are very valuable for muscular development, they are of still greater value for the hygienic effects which they produce.

No class of exercises contains so many variations nor such gentle and even steps of progression. To describe them all in this book would take too much space, and we will confine ourselves to the main forms, merely naming the others. For ease of description, they have been grouped together as, — sideways flexions; rotations; both movements combined: rotation with flexion forward and backward; and leg elevations.

A lesson usually contains two or more lateral trunk-movements as different from each other as possible. Besides, some easy exercise of this kind which has been done before can be used with other introductions in every lesson.

The simplest sideways flexion is the

Close St. T. Sidew. Flex. (Fig. 176). — Command, "*Feet — clóse! Trunk to the left (r.) — bend!*" The trunk is slowly bent to the side named, the hands glide on the legs, the knees

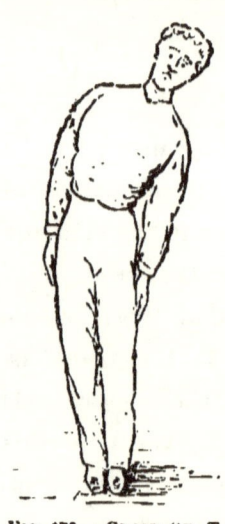

FIG. 176. — CLOSE ST. T. SIDEW. FLEX.

remain straight, and the head follows the movement of the shoulders. Common faults in all sideways flexions are: pushing one shoulder (in this case the right one) upward, and leaning the head against this shoulder, bending one knee, and bending the trunk forward or backward. The trunk is slowly straightened at the command, "*Upward — stretch!*" To increase the force of the movement and thus to get a progression out of one and the same exercise, the arms are subsequently placed in half str., rest, str., and yard *c* positions. Wing position is used but little in sideways flexions (except for children). If half str. pos. is used, the flexion takes place to the left, when the right arm is extended upward, and *vice versa*. Besides close st., the position of the feet can be: st., stride st., walk st. *b*, walk st. *c*, and fallout st. *b*. In the last three positions, the flexion takes place only to the side of the foot which is in front. For instance:

Str. Walk St. *b* **T. Sidew. Flex.** (Fig. 177). — Command, "*Left foot forward and arms upward — strétch! Trunk to the left — bend! Upward — stretch! . . . Arm-extension upward*

MOVEMENTS OF THE SWEDISH SYSTEM 153

and change of feet — óne! Twó! Trunk to the right — bend!" . . .

The most commonly used sideways flexions are:

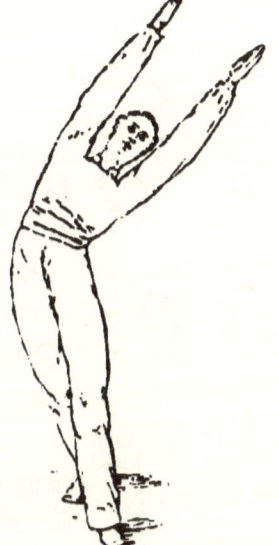

Fig. 177. — STR. WALK ST. b SIDEW. FLEX.

```
Close st.
Wing      ⎧ Close st.   ⎫
Half str. ⎪ St.         ⎪
Rest      ⎨ Stride st.  ⎬ T. sidew. flex.
Str.      ⎪ Walk st. b  ⎪
          ⎩ Walk st. c  ⎭
Rest ⎫
Str. ⎬ Fallout st. b
Yard stride st.
```

After the commencing position has been ordered, the command for the flexion is invariably as described above, and the execution is always slow.

Advanced pupils may do rest stride st. and str. stride st. T. sidew. flex. (Fig. 178) rapidly instead of slowly. Command. "*Left (r.) foot sideways place and neck — fírm!* (or *Arms upward — strétch!*) *Trunk-flexion quickly, left (r.), right (l.), and upward — óne! Twó! Thrée!*" or, later on. "*Trunk-flexion quickly left and right — óne! Twó! . . . Upward — stretch!*"

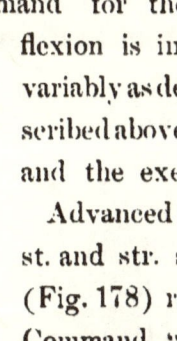

Fig. 178. — STR. STRIDE ST. SIDEW. FLEX.

Half Str. Half Wing Side Sup. Walk St. *b* **Sidew. Flex.** (Fig. 179). — The pupil stands beside the horizontal bar, which is put at hip height. He turns his side — supposing in this case the left — to the bar and stands so close that he touches the latter. Command, "*Left foot forward place, left hand hip firm and right arm upward stretch — óne! Twó!*" 1. The right arm bends. 2. The position named is taken. "*Trunk to the left — bend! Upward — stretch!*" . . . The movement, always in walk st. *b* pos., is also done with the arms in rest or str. pos.

Foot Side Gr. St. T. Sidew. Flex. (Fig. 180). — In the foot side gr. st. pos. the pupil stands with his side turned to the stall-bars and one step away from them. The foot nearest the bars is inserted between the second and third rounds, both knees are straight, and the body in usual good posture. Trunk-flexion takes place in direction from the bars, both knees remaining straight. The flexion should not be deeper than that the trunk and elevated leg are in a straight line. The command is as described above, and the movement is done with the arms in wing, rest, or str. pos. In absence of stall-bars the foot may be placed on the horizontal bar, or on a chair, or the knee of another

FIG. 179.— HALF STR. SIDE SUP. WALK ST. *b* SIDEW. FLEX.

FIG. 180.— STR. FOOT SIDE GR. ST. SIDEW. FLEX.

pupil (compare Fig. 165, p. 143), where the latter holds it in position.

Trunk Rotation. — In this movement the trunk is rotated around its own axis; the body below the waist remains as immovable as possible, and the head, shoulders, and arms follow the movement of the trunk. Common faults in this movement are pushing the shoulder forward, turning the head so that it loses its position relative to the shoulders, bending the trunk to the side or forward, bending the knee of the side toward which rotation takes place, and turning the feet out of place. The simplest movement of this kind is

Wg. Close St. T. Rot. — Command, "*Feet close and hips — firm! Trunk to the left (r.) — turn! Forward — turn!*" ... The movement is done slowly. When the trunk is rotated to either side, the position is called **turn st. pos.** The movement is made stronger by using stretch (Fig. 181) or rest pos. of the arms. In all trunk-rotations the str. pos. should be used earlier than the rest pos., as the former is less apt to produce faulty posture of the shoulders and head. For, in trunk-rotation, with neck firm, the tendency is to bring one elbow too far forward, and at the same time to push the head forward with the hands. The rotation is performed on the same principles in the st. and stride st. positions. In the walk st. positions (all of which may be used), the rotation takes place to the side of the foot which is in front. For instance,

FIG. 181. — STR. CLOSE ST. T. ROT.

Wg. Walk St. *b* **T. Rot.** (Fig. 182). — Command, "*Left (r.) foot forward and hips — firm! Trunk to the left (r.) turn! Forward — turn!* ... *Change feet — óne! Twó! Trunk to the right (l.) — turn!*" ...

For advanced classes, the rotation may be done also to the opposite side in the str. walk st. *c* pos.

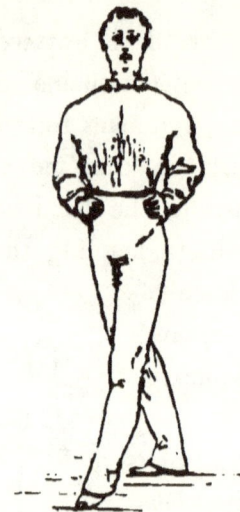

FIG. 182.— WG. WALK ST. *b* T. ROT.

Of fallout st. positions, only *a* is used, and here the rotation is done to either side. When the trunk is turned in direction of the forward foot, the position is called **turn fallout st.** *a* **pos.** (Fig. 183); and, if in the opposite direction, **reverse turn fallout st.** *a* **pos.** (Fig. 188). The last one can be taken so that the rotation and foot movement are done simultaneously. Command, "*Hips — firm! With rotation, to the left — fallout! Change feet and sides — óne! Twó!*" . . . When done in this manner, the head is turned in the direction of the forward foot. In half str. pos. this is a truly æsthetical movement: it is done as follows:

Half Str. Rev. Turn Fallout St. *a* **Pos.** (compare Fig. 145).—Command, "*Left arm upward, right arm backward, with rotation, to the left fallout — óne! Twó!*" 1. Bend st. pos. 2. Hlf. str. rev. turn fallout st. *a* pos.; the feet are in fallout st. *a* pos.; the trunk and backward leg are in a straight line: the trunk is rotated so that the shoulders are parallel with a line through the heels; the arms are

FIG. 183.— WG. TURN FALLOUT ST. *a* POS.

parallel with the backward leg,— the left has the palm turned up, the right the palm turned down; the head is turned in

direction of the left foot. "*Change arms, feet, and sides — óne! Twó!*" (As above.) An advanced class may do the change of feet zigzag forward (or backward). Command, " *Change arms, sides and feet forward (backward) — óne! Twó!*" ...
1. The backward foot is placed beside the forward one, and the arms bent. 2. The position is taken to the other side by moving this same foot sideways forward, etc. This form of the movement requires a great deal of muscular control, and should be used with moderation.

All the above rotations may be brought together as follows:

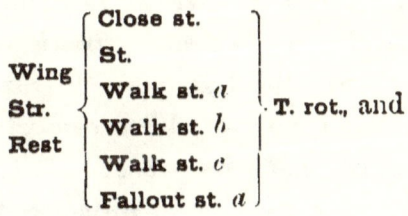

Fig. 184. — Str. Stride Sitt. T. Rot.

Wing Str. Rest { Close st. / St. / Walk st. *a* / Walk st. *b* / Walk st. *c* / Fallout st. *d* } T. rot., and

Half Str. Rev. Turn Fallout St. *a* Pos., with variations.

Children often find it difficult to keep their feet still while the trunk rotates. In such a case, let the pupils sit astride a bench (or chair) (Fig. 184), and command the movement in usual order. For children the rotation may also be done in knee st. pos.

In the str. stride st. pos., the rotation may be done by advanced pupils in quick time. The command and execution resemble those of the corresponding sidew. flex., and need no description here.

To increase the effect of a rotation, the pupils may be kept in turn pos. for a while; they are kept busy by a 2 A. ext. upw. For instance,

Str. Turn Stride St. 2 A. Ext. Upw. (Fig. 185). — Command, "*Feet sideways place and arms upward — strétch! Trunk to the left — turn! Arm-extension upward — óne! Two! . . . Trunk to the right — turn!*" etc. This movement is done in the positions given below:

Str. turn $\begin{cases} \text{Close st.} \\ \text{St.} \\ \text{Stride st.} \\ \text{Walk st. } a \end{cases}$ 2 A. ext. upw.

In walk st. *a* position, it may be changed for advanced pupils as follows: Command, "*Left foot sideways forward, to the left, and arms upward stretch — óne! Twó!*" 1. Bend st. pos. 2. Str. turn walk st. *a* pos. — "*With arm-extension, change feet and sides — óne! Twó!*" . . .

Half Yard Reach Turn Stride St. T. Rot. with 2 A. Fling. (Fig. 186). — This is a very difficult movement, used only for far-advanced classes. For the commencing position, command, "*Left foot sideways — pláce! Trunk to the left — turn! Left arm sideways, right arm forward — strétch! Right hand — túrn!*" (Palm up.) For the movement, "*Trunk-rotation quickly, with arm-flinging and turning of hands — óne! Twó!*" . . . 1. The trunk is quickly rotated to the right, while the forearms are slightly lifted, the hands turned and flung to the right, until the left arm comes into reach pos. with the palm up, the right into yard pos. with the palm down. 2. The movement is done on the same principles, to the other side. Before the movement reaches this complicated state, the rapid rotation should be practised alone in yard stride st. pos., and

FIG. 185. — BEND TURN STRIDE ST. 2 A. EXT.

the arm-flinging should be practised separately (as a shoulder-blade movement).

If the rotation is to be combined with sideways flexion, the flexion is done to the side to which the trunk is rotated. The commands are, for instance, for

Str. Turn Stride St. T. Sidew. Flex. (Fig. 187). — "*Left foot sideways place and arms upward — strétch! Trunk to the left*

FIG. 186. — HALF YARD REACH TURN STRIDE ST. T. ROT. WITH 2 A. FLING.

— *turn! To the left — bend! Upward — stretch! . . . To the right — turn!*" . . .

This movement is done in various positions, viz.:

$$\left.\begin{matrix}\text{Wing} \\ \text{Rest} \\ \text{Str.}\end{matrix}\right\} \text{turn} \left\{\begin{matrix}\text{Close st.} \\ \text{St.} \\ \text{Walk st. } a \\ \text{Walk st. } b\end{matrix}\right\} \text{Sidew. flex.}$$

$$\left.\begin{matrix}\text{Rest} \\ \text{Str.}\end{matrix}\right\} \text{Rev. turn fallout st. } a \text{ (Fig. 188)}$$

These will hardly need any description. The most common fault in this movement is bending the trunk forward instead of sideways.

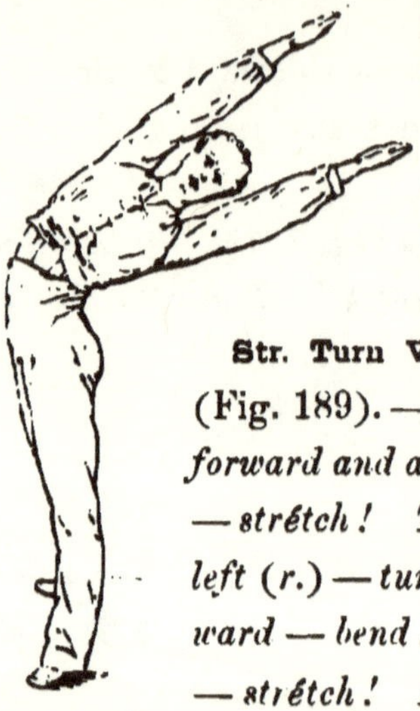

In the turn st. pos., flexion can also be done backward and forward in direction of the face (or neck). For instance:

Str. Turn Walk st. *a* **T. Flex. Backw. and Forw.** (Fig. 189). — Command, "*Left (r.) foot sideways forward and arms upward — stretch! Trunk to the left (r.) — turn! Backward — bend! Upward — stretch! Forward — bend!*" . . . (see Fig. 190). This movement can be made

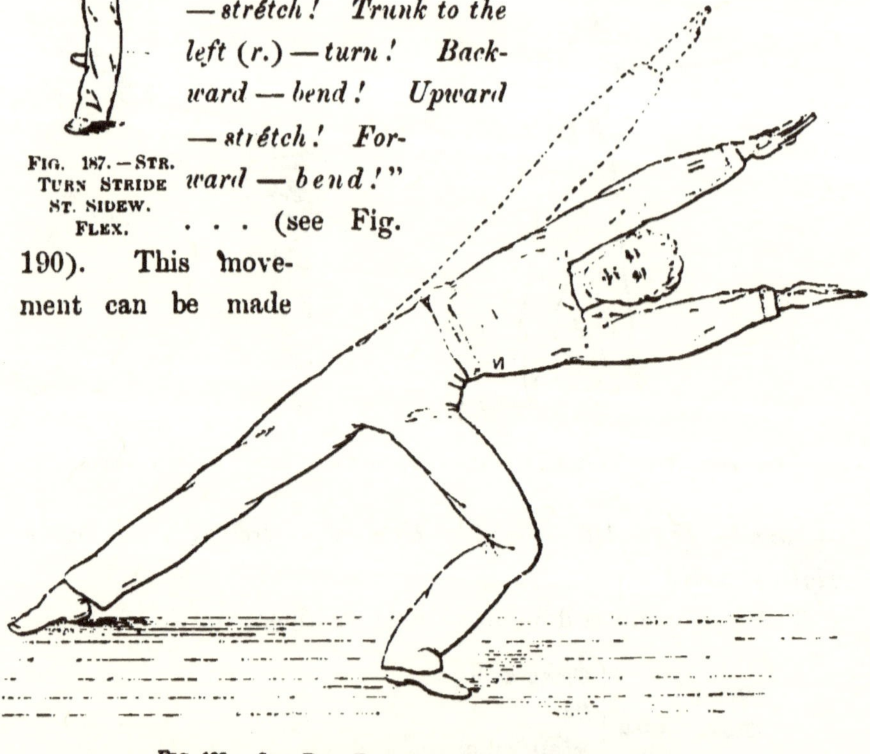

FIG. 187. — STR. TURN STRIDE ST. SIDEW. FLEX.

FIG. 188. — STR. REV. TURN FALLOUT ST. *a* SIDEW. FLEX.

exceedingly strong if the pupils remain in the backward flexion and execute arm-extension upward with change of feet and rotation to the opposite side. The movement is then called

Str. Arch Turn Walk *a* **St. 2 A. Ext. and Change of Feet.**— After backward flexion is taken, command, "*Arm-extension, change feet and sides — One! Two!*" ... This movement can be done only by very strong persons. At first the 2 A. ext. is done without change of feet.

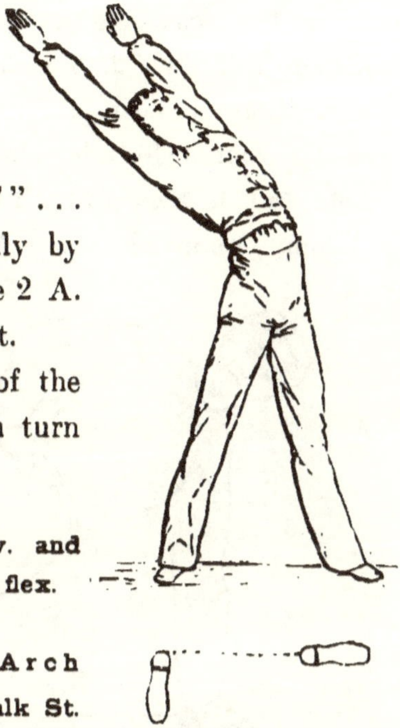

The following is a synopsis of the backward and forward flexions in turn positions:

Wing
Rest } turn { Close st.
Str.   St.   } Backw. and
              Walk st. *a* | forw. flex.
              Walk st. *b*

Str. Arch Turn Walk St. *a*, **2 A. Ext. and Change of Feet.**— This form of lateral trunk-movement resembles closely the arch-flexions (oblique), and may be used as substitutes for these when there are no bar-stalls or other suitable apparatus at hand.

FIG. 189.— STR. TURN ARCH WALK ST. *a* POS.

FIG. 190.— WING TURN ST. FORW. FLEX.

Side Fall. Pos. (Fig. 191).—To take this position, command, "*Side falling position, on the left hand — One! Two! Three!*" 1 and 2 are executed as stoop falling position (Figs. 167 and 168, p. 145). 3. The right hand takes "hip firm," while the body is turned over (to the right), so that it rests on the left hand and on the outside of the left foot. The feet are in close pos. and the legs straight. *Change hands — One! Two!"* . . . 1. Stoop fall. pos. 2. Side fall. pos. on the other hand. After this position has been practised for some time it is used as commencing position for a leg-elevation.

Side Fall. L. Elev. (Fig. 191). — After commencing position is taken, command, *"Leg-elevation — one! Two!"* . . . The

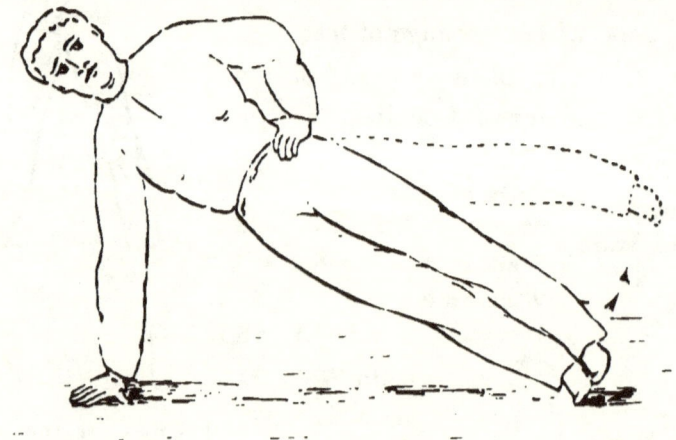

FIG. 191. — WING SIDE FALL L. ELEV.

upper straight leg is lifted as high as good posture will allow. The movement is also done in half str. side fall. pos. (Fig. 192). If the arm which is extended upward grasps a stall-bar (hor. bar, the hand of another pupil, etc.), the position is called half str. gr. side fall. pos., which is naturally easier to maintain than the same free position. The three movements thus received are:

        Side fall.⎫
        Hlf. str. gr. side fall. ⎬ **L. elev.**
        Hlf. str. side fall. ⎭

MOVEMENTS OF THE SWEDISH SYSTEM

Str. Side Sup. St. Sidew. Flex. and L. Elev. (Fig. 193). — The pupil stands close to the horizontal bar, put at hip height, with his side turned to the bar and touching it. Command, "*Arms upward — strétch! Trunk to the left (r.) — bend!*" The pupil bends over the bar, keeps the leg nearest it straight, and lets the other rise as the trunk bends, so that this leg and the trunk are in a straight line. "*Upward — stretch!*" The trunk is raised, and the elevated leg again touches the floor.

Str. Side Gr. St. L. Elev. (Fig. 194). — The pupil stands one step away from the stall-bars, his side turned to them. Command, "*Arms upward — strétch! Trunk to the left (r.) — bend! Grásp!*" The trunk is bent toward the bars, and each hand grasps a bar, so that the arms are straight (the upper

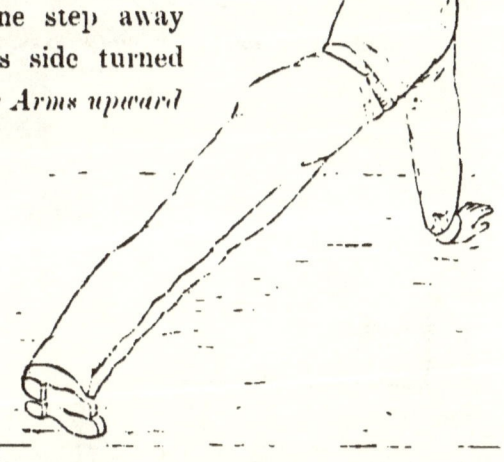

FIG. 192. — HALF STR. SIDE FALL. POS.

arm remains in str. pos., whereas the lower, to become straight, must be lowered into yard pos.; the palm of this hand is turned up). The leg nearest the bars is vertical and straight. The foot of the other leg is raised above the floor. "*Leg-elevation — óne! Twó!*" ... The outside leg is raised and lowered.

Introduction to Wheeling (Fig. 195). — The pupil stands one step away from a bench, with his side turned to it. Command, "*Right (l.) foot sideways and arms upward — strétch!*" Rapid

164   *EDUCATIONAL GYMNASTICS*

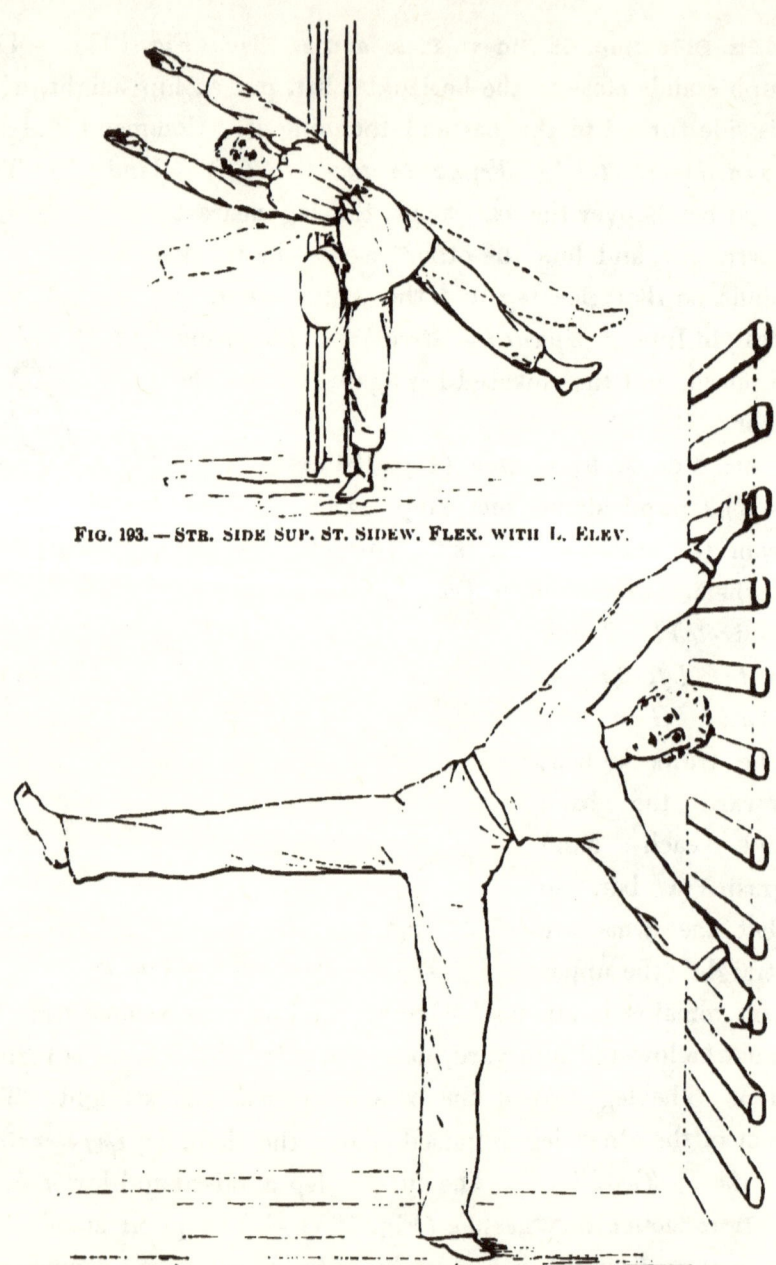

Fig. 193. — Str. Side Sup. St. Sidew. Flex. with L. Elev.

Fig. 194. — Str. Side Gr. St. L. Elev.

trunk-flexion to the left (*r.*) — *óne! Twó!*" . . . 1. The pupil bends to the left (r.); at the same time he raises his right (l.) leg, and lowers his left (r.) arm until his left (r.) hand rests on the bench (the position is like str. side gr. st. L. elev.). 2. He resumes str. stride st. pos.

Wheeling (Fig. 196). — Command, "*Feet sideways place and arms upward — strétch! To the left* (*r.*) — *whéel!*" The pupil bends his trunk slightly to the right (l.), and then with utmost speed to the left (r.), until his left (r.) hand rests on the floor; his right (l.) hand is placed on the floor beyond the left (r.) one; his feet, having left the floor, describe a semi-circle in the air, and land on the floor beyond his right (l.) hand. By this time his hands have left the floor, and he stands in str. stride st. pos. The motions of arms and legs resemble the spokes of a revolving wheel.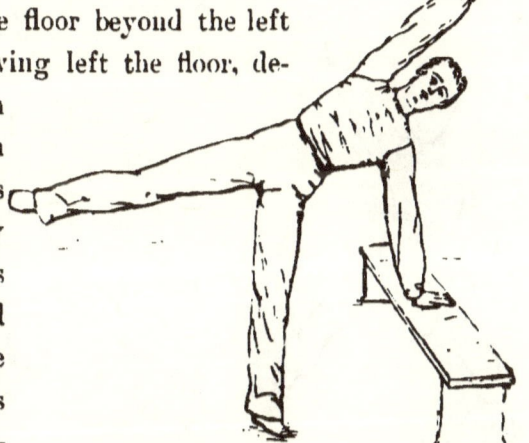

FIG. 195. — INTRODUCTION TO WHEELING.

The hands should not touch the floor simultaneously, nor should the feet, but the movement should be a distinct "one — two — three — four." Legs as well as arms are kept straight throughout the movement. When practised for the first time, the pupils may be allowed to bend slightly forward, and, already in the commencing position, to lower the hand in whose direction the wheeling is to be done. As soon as every pupil has gained some proficiency, the class is arranged in one rank, the pupils standing behind each other

166   EDUCATIONAL GYMNASTICS

at "whole distance." After they have been numbered by twos from the front backward, command, "*Feet sideways place and arms upward — stretch! Numbers one to the left, numbers two to the right — wheel!*" The ones and twos do the movement in opposite directions, as ordered. "*To the opposite side — wheel!*" The pupils wheel back into their original places.

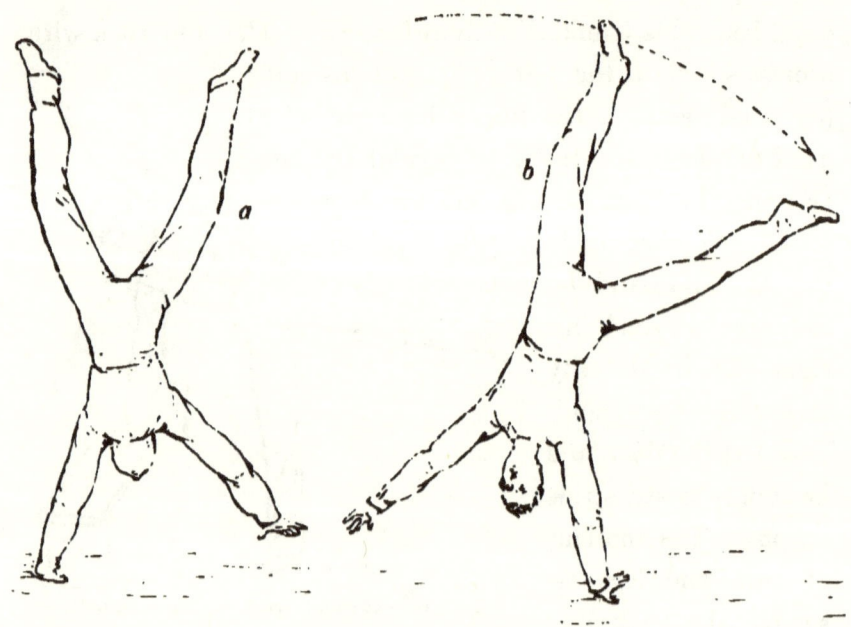

FIG. 196. — WHEELING.

The movement must take place strictly sideways, or accidents will happen (the gymnasts interfere with each other); consequently wheeling must not be tried in class until after considerable individual practice.

Sidew. Hang. Pos. (Fig. 197). — The pupil stands at the barstall with his side turned toward the wall and one step away from it. Command, "*Arms upward — stretch! Trunk to the*

MOVEMENTS OF THE SWEDISH SYSTEM 167

left (r.) —*bend! Grásp!* (see Fig. 194.) *Sideways — hang óut!"* The pupil swings both his legs up, and holds himself out from the bars by means of his arms alone. Arms and legs are straight, and the body stands out at right angles from the bars. The pupil stays in this position as long as his strength will allow, and then lowers himself slowly into commencing position. Beginners may learn to take this position by first taking the str. side gr. st. pos. with elevated outside leg. This leg is now lowered with utmost speed (and allowed to pass in front of and on the other side of the inside leg), and immediately swung back again as high as possible, the inside leg following close to it. The upper arm may at the same time be bent; and, as the body is lowered into horizontal position, it is straightened. Advanced pupils get into the position by merely elevating their legs with moderate speed until they are horizontal. This, however, requires a high degree of physical development. A common fault in this exercise is turning the face obliquely upward, and curving the body and legs (Fig. 198); in fact, there are but few who reach the ideal in this position.

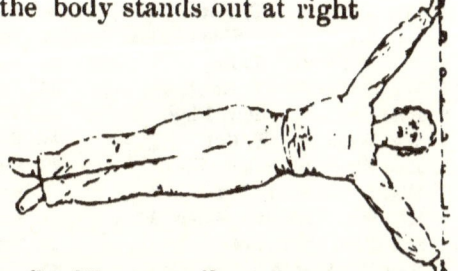

FIG. 197.— SIDEW. HANG. POS. AS IT OUGHT TO BE.

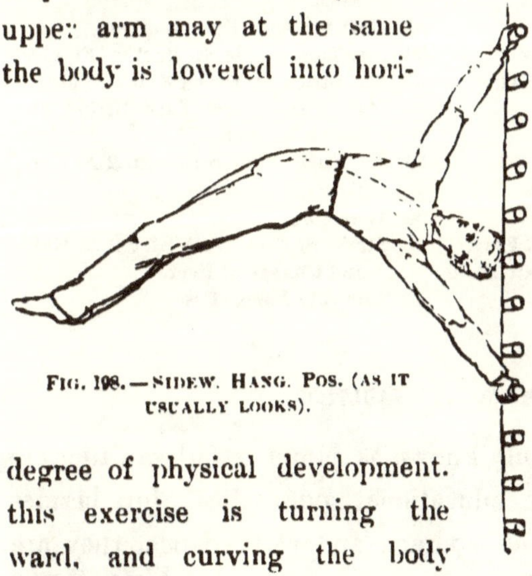

FIG. 198.— SIDEW. HANG. POS. (AS IT USUALLY LOOKS).

Progressive arrangement of lateral trunk-movements:

1. Close St. Sidew. Flex.
2. Wing Close St. T. Rot.
3. Wing Stride St. Sidew. Flex.
4. Wg. Walk St. a T. Rot.
5. Hlf. Str. Close St. Sidew. Flex.
6. Wg. Walk St. b T. Rot.
[7. Wg. Stride Sitt. T. Rot.]
8. Rest Close St. Sidew. Flex.
9. Wg. Walk St. c T. Rot.
10. Rest Stride St. Sidew. Flex.
11. Wg. Stride St. T. Rot.
12. Hlf. Str. Walk St. c Sidew. Flex
13. Str. Close St. T. Rot.
14. Rest Walk St. b Sidew. Flex.
15. Bend Turn St. 2 A. Ext. Upw.
[16. Str. Stride Sitt. T. Rot.]
17. Wg. Close St. T. Backw and Forw. Flex.
18 Str. St. Sidew. Flex.
19. Str. Walk St. a T. Rot.
[20. Str. Kn. St T. Rot.]
21. Rest Turn Close St. Sidew. Flex.
22. Str. Turn Walk St. a 2 A. Ext. Upw.
23. Rest Foot Gr. Half St. Sidew. Flex.
24. Str. Stride St. T. Rot.
25. Rest Turn Walk St. a Sidew. Flex.
26. Bend Turn Stride St. 2 A. Ext.
27. Str. Stride St. Sidew. Flex.
28. Wing Fallout St. a T. Rot.
29. Rest Turn Close St. Backw. Flex.
30. Str. Walk St. b T. Rot.
31. Str. Walk St. b Sidew. Flex.
32. Rest Turn Stride St. Sidew. Flex.
33. Half Str. Rev. Turn Fallout St. a Pos.
34. Str. Turn Walk St. a 2 A. Ext. and Change of Feet.
35. Str. Turn Walk St. a Backw. Flex.
36. Str. Walk St. c T. Rot.
37. Str. Side Sup. Walk St. b Sidew. Flex.
38. Str. Turn St. Backw. Flex.
39. Str. Foot Side Gr. Half St. Sidew. Flex.
40. Wg. Side Fall. L. Elev.
41. Rest Fallout St. a T. Rot.
42. Str. Turn Stride St. Sidew. Flex.
43. Str. Turn Close St. Backw. Flex.
44. Str. Side Gr. Half St. L. Elev.
45. Str. Fallout St. a T. Rot.
46. Hlf. Str. Gr. Side Fall. L. Elev.
47. Str. Walk St. c Rev. T. Rot.
48. Str. Side Sup. St. Sidew. Flex. and L. Elev.
49. Str. Stride St. Quick T. Rot.
50. Str. Walk St. b Backw. Flex.
51. Str. Turn Walk St. a Sidew. Flex.
52. Yard Stride St. Sidew. Flex.
53. Str. Turn Arch Walk St. a 2 A. Ext.
54. Str. Rev Turn Fallout St. a Sidew. Flex.
55. Half Str. Side Fall. L. Elev.
56. Str. Turn Stride St. Backw. Flex.
57. Str. Stride St. Quick Sidew. Flex.
58. Half Yard Reach Turn Stride St. Quick T. Rot.
59. Str. Turn Arch Stride St. 2 A. Ext. Upw.
60. Wheeling.
61. Str. Turn Arch Walk St. a 2 A. Ext. and Change of Feet.
62. Sideways Hang. Pos.

JUMPING AND VAULTING.

The exercises commonly known as jumping and vaulting are of the greatest value in educational gymnastics; for, besides their effect of developing courage and self-reliance, they are often of great practical use in ordinary, every-day life. But still more important is their effect of promoting the general elasticity of the body. A person may be muscularly strong and yet not be able to use his strength to best advantage simply for want of elasticity — want of grace; for, elasticity and grace are, in a measure, one and the same thing. There are no other exercises which require so much general muscular control — co-ordination of movements — as do jumping and vaulting; to do these well, it is necessary to be truly "well-balanced"; and for this reason it may be said that the degree of perfection with which a person jumps and vaults may usually be considered as an indicator of that person's physical development (in the sense in which we use this word).

No movements need a more thorough preparation, and none will so soon show bad results if they are applied in wrong progression, as these jumping and vaulting exercises: and yet in these movements no system of gymnastics, except the Swedish, has any progression worthy of the name.

In jumping, the question is less how high to jump, more in what manner is it to be done. Many persons have the idea that, after the feet leave the ground, the movement is ended, and that it matters little whether the landing is made on the feet or on the head, as long as the mattress is there to receive them. The most important part of jumping, however, is the manner in which the landing should be done; for, unless the fall is properly broken by the body itself, jumping loses all practical value, since mattresses are not to be found out of doors.

The knee-joint is so shaped that, when the knee is straight, the surfaces of the condyles fit very accurately into the facets of the head of the tibia; and the first part of a knee-flexion takes place slowly. The radius of the condyle is decreasing backward, so that, after flexion has once begun, it occurs easily and with accelerating speed. Now, in jumping, if the landing is made with straight knees, the facets of the knee-joint are pressed forcibly together, so that flexion becomes impossible, and the thighs and hips suddenly oppose the downward movement; but gravitation cannot be thus suddenly overcome, and the body still remains for a while under the influence of the fall; the neck of the femur has to resist the pressure from above, and, if the speed of motion is great, the bone will break; the internal organs also continue to move downward, and the ligaments, etc., by which they are suspended, will thus be excessively stretched, until the fall is broken — displacement is often the result. On the other hand, if the knees have already commenced to bend, when the feet strike the ground, they continue to do so until the fall is wholly broken, and the motion is gradually reversed through the elasticity of the quadriceps extensor, etc. For ordinary height and speed, all danger of fracture and displacement is now removed. The heels should be raised, when the feet strike the ground, so that the gastrocnemius may also help to break the fall. Furthermore, the heels should be held together and the feet turned out, for, in this position, the neck of the femur acts like an oblique beam, distributing the pressure throughout the whole bone, whereas, when the heels are apart, the neck is horizontal, and hence easily broken. Besides, the position lessens the danger of internal displacement. If done in this way, there is no need of a mattress to land on when ordinary jumping is practised. In the more violent forms of vaulting, and when jumping takes place from great heights, a mattress may be used as a safeguard against accidents and to prevent the feet from becoming sore. On any other occasion, the use of a mattress will tend to ruin the form of the movement and to take away its best effects.

This class of exercises contains such an infinite variety that it will be possible to describe only the most essential forms; the teacher easily supplying the others, if he has need of them.

Preparation to Jumping (Fig. 36, p. 56). — Command, "*Prepare to jump — óne! . . . Fóur!*" . . . The execution is like that of st. 2 Kn. flex. (see leg-movements). At first the movement is practised with hips firm, later on without it, the arms hanging straight behind the thighs when the knees bend. From time to time the teacher changes the rhythm of this movement ("*— one͡ twó! — Thrée! — Fóur!*" or "*— óne! - Two͡ thrée! — Four!*" etc.), so that the pupils become thoroughly balanced in each one of the intermediate positions. After some time, the teacher leaves off counting, and merely commands, "*Preparation to jumping — stárt!*" when the pupils go through the whole movement. When it is done in this way, the teacher may occasionally and unexpectedly say, "*Twó!*" (or "*Thrée!*" etc.), when the pupils understand that they are to stop in the position belonging to that count. This is an excellent way to make the pupils well balanced in the intermediate positions. After this movement can be well performed, it is safe to proceed to the next one; yet the preparation to jumping should be brought in in almost every lesson; for, no matter how skilled the pupils may be, the movement always retains its effect of correcting the form of all jumping and vaulting. It may then be used as an introduction or as a slow leg-movement, since it shares the effects of the latter kind of exercises, even though it is done in rapid rhythm.

Upw. Jump (Fig. 199). — Command. "*Upward jump — óne! Twó! Thrée! Four! Five!*" 1. Heel elev. 2. 2 Kn.

flex. to 90°. 3. By a rapid extension of his knees, the pupil jumps straight up; and, in the same place where he stood before, he lands on tip-toe with raised heels, and lets his knees bend to right angles; in this position he stays with erect trunk and head, and arms hanging straight behind his thighs. 4. The knees are stretched. 5. The heels are lowered. This movement should never be done with "hips firm;" for what is gained by the knee-flexion in landing is apt to be lost by the pressure exerted by the hands on the hips.

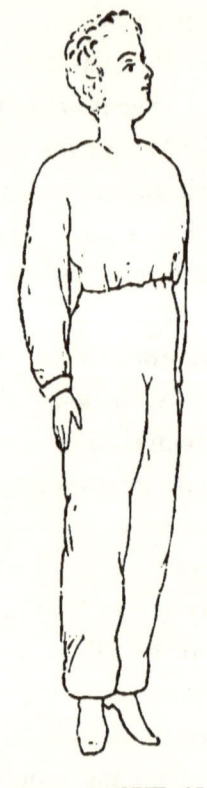

Fig. 199. — Upw. Jump. "Three!"

Turn. 90° Upw. Jump. — Command, "*Turning 90° to the left (r.), upward jump — óne! . . . Five!*" The movement is executed as the one just described, except that, as the pupil jumps, he turns 90° to the side ordered, and faces in this new direction when he lands. [Later on 180° (and even 360°) may be tried.]

With A. Fling. Upw. Jump. — Command, "*Arms flung sideways, upward jump — óne! . . . Five!*" This resembles upward jump; but, as the pupil jumps, he flings his arms straight sideways to horizontal position, thus aiding the motion upward. The arms move quickly down again, so that they are in the usual position when the landing occurs.

With A. and L. Fling. Upw. Jump. — Command, "*Arms and legs flung sideways, upward jump — óne! . . . Five!*" At the jumping, arms and legs are flung out sideways, and brought back into position again, so that the arms are perpendicular, and the heels together, when the feet strike the ground.

Forw. Jump (Fig. 200 and 201). — Command, " *Forward jump — óne!* ... *Five!* " 1. Heel-clev. 2. 2 Kn. flex., and the arms are moved backward into backw. str. pos. 3. Keeping his heels together, the pupil jumps forward, and helps the movement by swinging his arms straight forward into reach pos. The arms swing back again; and, by keeping them rigid behind the thighs when the landing occurs, the body is prevented from tipping over forward. 4. Knees stretch. 5. Heels sink.

FIG. 200. — FORW. JUMP: THE START.

Backw. Jump. — On the same principles, jumping can be done backward; this movement, however, is of less value, and is but little used.

Sidew. Jump. — Command, " *To the left (r.) jump — óne!* ... *Five!* " This is executed on the same principles as forward jump, except that, just before jumping, the pupil swings his straight arms in front of him to the side opposite the one to which he is jumping; and, as he jumps straight to the side, the arms are flung in direction of the jumping, and brought into the usual position at the landing so as to prevent the trunk from tipping over.

Later upw. jump. may be practised with a command for only the first and last part of the movement. Command, " *Upward —jump! Five!*" The pupils go through the first four parts with utmost speed. Before this, however, the movement should be done as, "*Upward—jump! Four! Five!*" the pupils stopping at the landing.

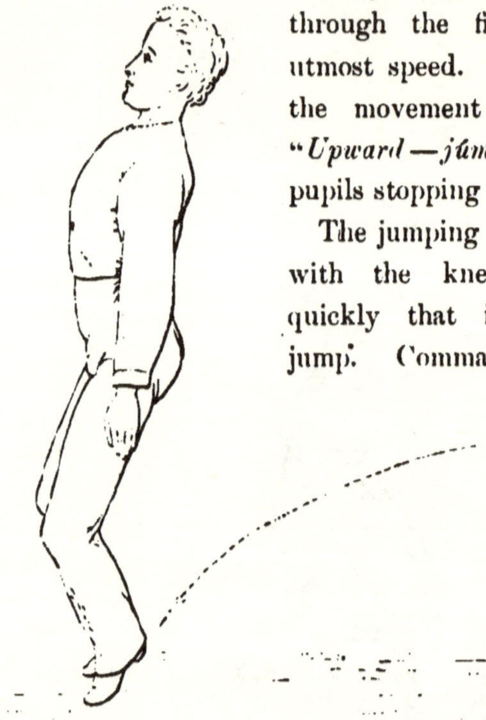

The jumping may also be practised with the knee-extension done so quickly that it creates one more jump. Command, "*Twice upward jump — one! . . . Six!*" or, "*Turning 90° (180°) to the left (r.) and back to the right (l.), twice upward jump — one! . . . Six!*" or, "*To the left (r.) and right (l.) jump — one! . . . Six!*"

FIG. 201.— FORW. JUMP: THE LANDING.

If the start is given by both feet, heels together (as in the previous exercises), it is called "whole start;" if by only one foot, "half start." In every-day life, the most common form of jumping is done with "half start," one foot giving the spring, as we walk or run, the other merely oscillating forward. The following exercises are instances of this form:

MOVEMENTS OF THE SWEDISH SYSTEM 175

Forw. Jump w. One Step's Start. (Fig. 202). — Command, "*Left (r.) foot forward, forward jump — one⌢two!! Three! Four!*" 1. The left (r.) foot is placed forward in walk st. b. pos. 2. (Follows instantly.) The right (l.) leg and both arms are swung forward with utmost speed, while the left foot springs from the ground, and the body is thrown forward; the heels are brought together, and landing takes place in usual form. 3. Knees stretch. 4. Heels sink. The movement is

FIG. 202. — FORW. JUMP. WITH ONE STEP'S START.

repeated on each foot alternately, so that the pupils become equally skilled in using either for the start. At first the distance jumped should be small, so that the landing may not lose its correct form. Later on, two or three steps are used for the start. For instance,

Forw. Jump w. 3 Steps' Start. — Command, "*Three steps' start; begin with the left (r.) foot, forward jump — one⌢two⌢thrée!! Four! Five!*" (or -- "*stárt! Four! Five!*"). The third step is the "half start" for the jumping, executed as above. After three steps have once been used, the movement is always done in this way, as one or two steps do not produce the speed that should characterize all jumping.

This form of jumping may also be done with 90° turning; the turning is then to the side of the foot which gives the final start. For instance,

90° Forw. Jump w. 3 Steps. — Command, "*With 90° turning and three steps' start, beginning with the left (r.) foot, forward —júmp! Four! Five!*" The pupils face to the left (r.) when they land. If the start had been with two steps, they would have faced to the right (l.), etc.

The jumping may also be done sideways forward, and then always with one step's start:

Sidew. Forw. Jump. — Command, "*Forward to the left (r.) jump — one two! Thrée! Four!*" 1. The right (l.) foot is placed crosswise forward to give the start. 2. (Follows instantly). By swinging the left (r.) foot in its own direction and sending the right (l.) foot from the ground, jumping takes place sideways forward to the left (r.); landing is done as usual (with heels together, etc.). 3. Knees stretch. 4. Heels sink. This form of jumping is one of the most graceful movements known.

In jumping with "half start," — running free jump, — the teacher soon leaves off counting the separate movements, and commands only the start and the last two parts (knee-extension, and lowering of the heels).

When proficiency is gained in these movements, two marks are made on the floor to indicate the distance to be jumped; and a rope is used to mark off the height for high jump. The distance or height should not be so great that the jumping cannot be done in correct form. When jumping over a rope (or other obstacle), the pupil should jump straight forward —

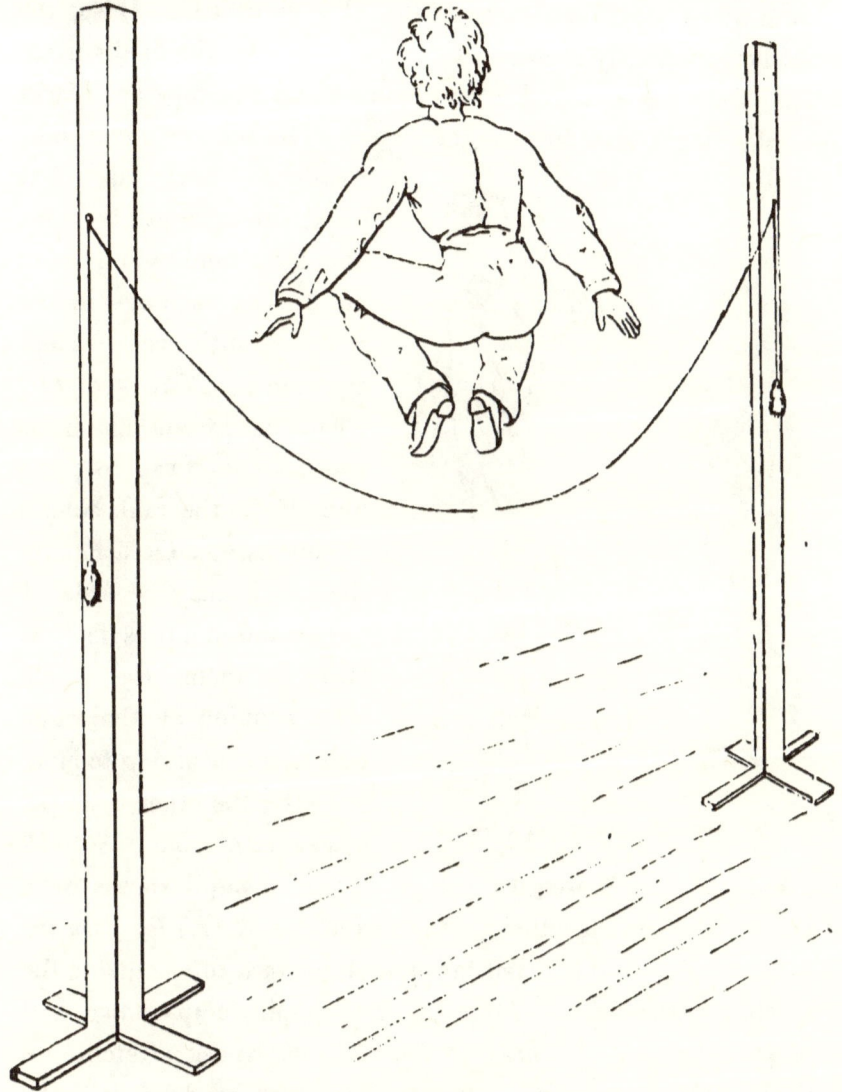

FIG. 203. — RUNNING HIGH JUMP.

not with legs swinging to one side — and bring his heels together already before passing the rope (Fig. 203). The jumping may be done either with whole or with half start; in the latter case not less than three steps are used; but usually more steps are taken in the form of a short run. If so, the teacher previously decides which foot is to give the final spring. The pupils are arranged in one rank behind the rope (or marks on the floor), and face to the flank. The teacher commands,

FIG. 204.—JUMPING FROM A BENCH.

"*Stárt!*" and the first pupil runs, jumps the rope, and, after landing, stretches his knees; he stays in toe st. pos. until the teacher commands, "*Next — stárt!*" when he lowers his heels and marches away to place himself in the rank behind the others, etc. Occasionally the pupils may be ordered to run and jump as fast as there is room, the whole class running at the same time and continuing to jump one after the other until the teacher commands, "*Stóp!*"

Jumping from a Height (Fig. 204).—The pupil stands on a bench (or other elevation). Command, "*Left (r.) leg forward — lift!*" The leg is lifted far enough forward to be outside the bench. "*Downward — júmp!*" The pupil jumps down and lands as usual. "*Thrée! Four!*" 3. Knees stretch. 4. Heels sink. The pupil may be allowed to bend slightly the knee of the leg which gives the start, especially if the height be great. This slight knee-flexion just before jumping also gives more grace to the movement. The pupil must not look down when jumping; for, if he does, he loses his balance when landing.

Star Gr. St. Jump from Stall-Bars (Fig. 205). — The pupil stands with one foot on the fourth or fifth bar, grasps with the hand of the same side a bar above his head, and holds himself out sideways, the other arm and leg being elevated sideways, so that the position of the limbs resembles the points of a star. Command, "*Downward jump — one ⁓ two !: Thrée ! Four !*" 1. The free arm and leg are swung in to the body. 2. They are again swung sideways, and, at the same time, the other foot and hand leave the bars, and the pupil jumps down sideways and lands as usual (facing as before jumping). 3. Knees stretch. 4. Heels sink.

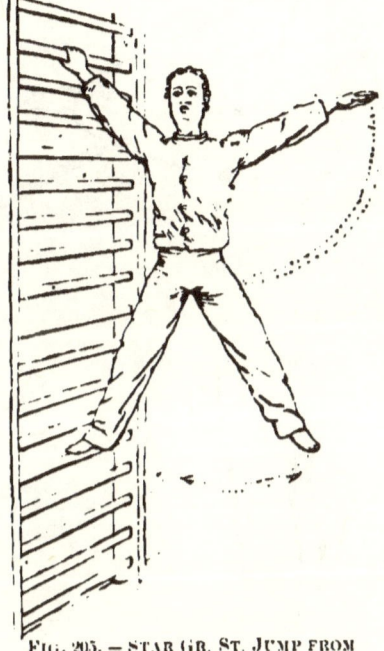

FIG. 205. — STAR GR. ST. JUMP FROM STALL-BARS.

Reach Gr. St. Introd. to Vaulting (Figs. 206 and 207). — The pupil stands at the horizontal bar, at whole distance from it. At the command, "*Grásp !*" he grasps the bar, thumbs behind, fingers in front (unless the bar be too thick to allow this). At the command, "*Sit up — one ⁓ two !*" he jumps and pulls himself forward, so that he lands against the bar; and, resting over it, he holds himself there, his arms straight, his head high, chest well expanded, and the body and legs forming a graceful curve backward (balance hang. pos.). At the command,

"*Three ͡ fóur!*" he swings himself off from the bar, his hands giving a strong push, and lands, in the usual manner, in the same place from which he started. After the first day, the teacher omits the command, "*Grásp!*" and the pupil does not place his hands on the bar until just before he jumps.

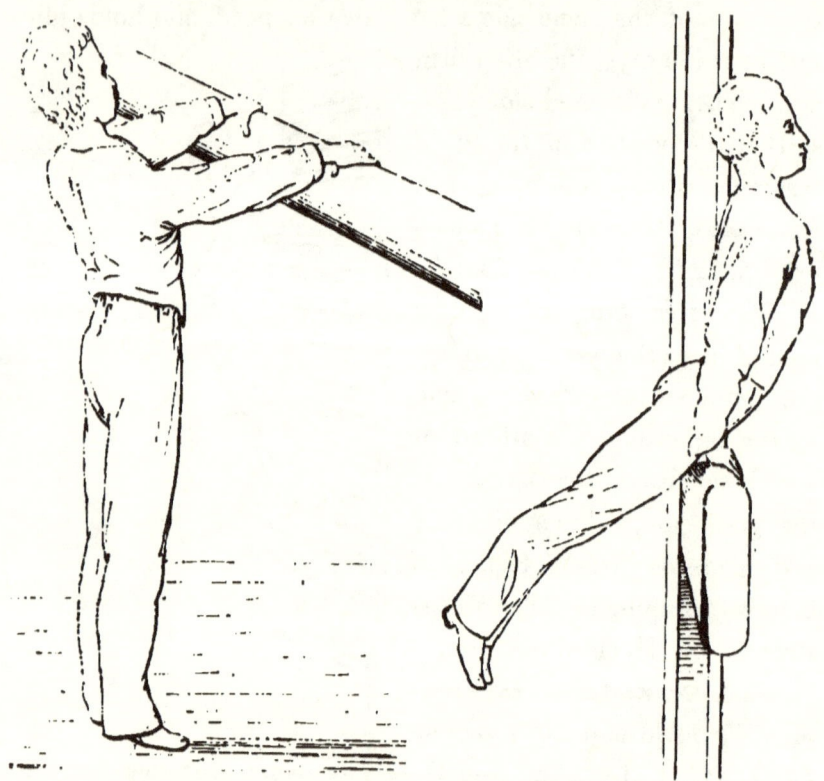

FIG. 206. — INTROD. TO VAULTING: THE MOMENT BEFORE JUMPING.

FIG. 207. — INTROD. TO VAULTING: "TWO." BAL. HANG. POS.

Half Yd. Gr. St. Introd. to Vault. (Fig. 208). — The pupil stands with his side turned to the bar, and grasps it with one hand (straight arm). At the command, "*Sit up — one ͡ twó!*" he turns toward the bar, grasps with the other hand also, and sits up as in the previous exercise. When jumping down, he pushes himself off by the forward hand, so that he lands in the same place and position (relative to the bar) which he had

Fig. 208. — Half Yd. Gr. St. Introd. to Vaulting. Commencing and Final Pos.

before sitting up. In these two exercises, the bar is at first placed at hip height and gradually raised as the pupils gain proficiency. As many pupils as the bar will hold do the movement at the same time.

If the bar is provided with saddles (or if there is a vaulting-horse or box), the following movement may be done:

Half Yd. Gr. St. Sit up Astride. — The pupil stands as in the previous exercise, and places his hand on one pommel of the saddle. At the command, "*Sit up — one͡ two͡!*" he sits up as in the previous exercise, one hand on each pommel. "*Three!*" He swings the leg, which in commencing position was outside, over the pommel, and sits gently down in the saddle as if on horseback; his legs and arms hang straight down, and his trunk and head are erect. At the command, "*Sit off — four͡ five͡!*" he grasps the pommel in front of him with both hands, swings his legs backward and over to the side from which he sat up; and, as he lands, he grasps the backward pommel with one hand, and faces the same way as he did when sitting in the saddle.

Introd. to Somersault over the Bar (Fig. 209). — The pupil stands facing the bar, and sits up into balance hang. pos., at the command as above. "*Hands — turn!*" He turns his hands, one at a time, so that the thumbs come in front, the fingers behind. "*Introduction to somersault — one! Two!*" . . . 1. The pupil folds himself forward over the bar, so that he hangs, head down, and holds himself in this position by means of his hands and thighs; his legs are bent at the hips and knees. 2. He pulls himself back into balance hang. pos. When this has been practised for a day or two, the somersault may be tried.

Somersault over the Bar (Fig. 210). — Command, "*Sit up — one͡ two! Hands — turn!*" (As above.) "*Somersault — three͡ four!*" The pupil folds over the bar, as in the previous exercise; and, turning his trunk fully under it, he lets his legs

MOVEMENTS OF THE SWEDISH SYSTEM 183

slowly swing over backward until they rest on the floor, so that he is in stride fall hang. pos. (Fig. 78). "*Five!*" He swings his body forward from under the bar, and, pushing well off with his hands, he rises into str. stride st. pos. "*Six!*"

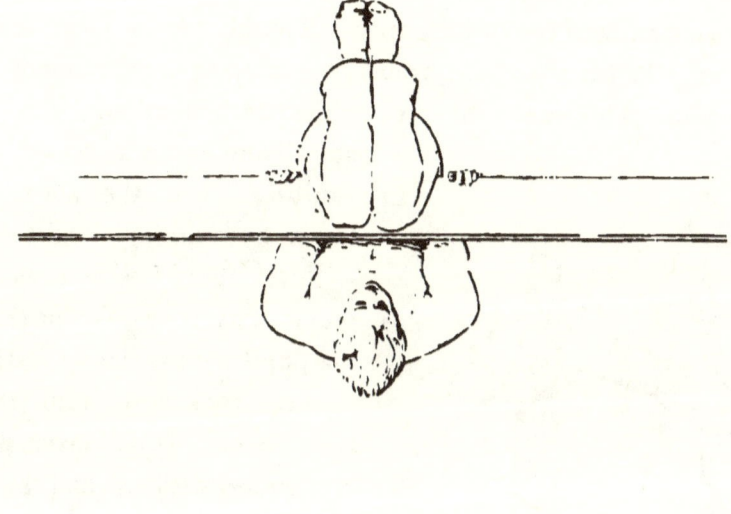

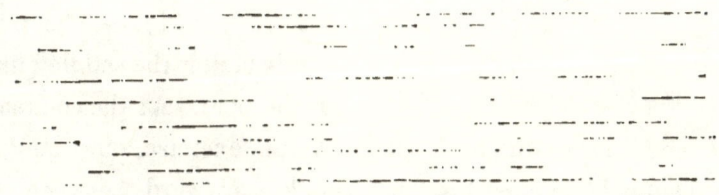

FIG. 209. — INTERMEDIATE POSITION OF SOMERSAULT OVER THE BAR.

He takes fundamental position. The corresponding arch-flexion serves as an introduction to this movement. The movement is made more difficult by placing the bar nearer the floor, until finally there is barely room for the trunk between the bar and the floor. When the bar is low, the hands are placed in turned position, and the sitting up omitted.

Walk St. *b* Introd. to Vaulting. — The pupil stands a full step away from the bar and facing it. Command, "*Left* (*r.*) *foot backward — pláce! Sit up — one˘twó!*" (At first . . . — "óne! Twó!") 1. The pupil places his left (r.) foot forward, and springs as in "Left (r.) foot forward, forward jump!" 2. He lands on both feet, which are put forcibly on the floor; and, grasping the bar with both hands, he sits up as in the previous exercises. This mode of giving the start is called "double start," and is always used in vaulting when the start is taken by running. "Sit off — three˘fóur!" is done as in previous exercises, except that the pupil throws himself still farther backward from the bar. Later on, this movement is practised with two and three steps' start.

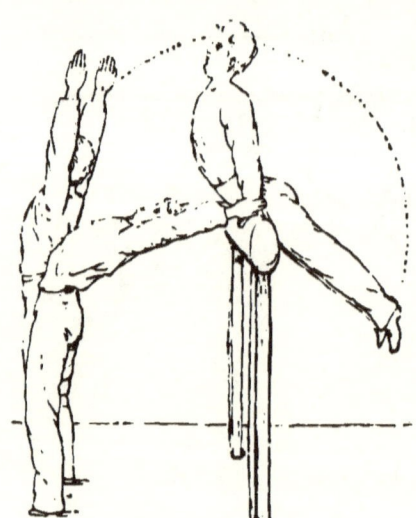

FIG. 210. — SOMERSAULT OVER THE BAR.

Vaulting between the Pommels (Fig. 211). — The pupil stands facing the saddle (fixed on the bar). At the command "*Spring!*" he sits up as in the previous exercise, one hand on each pommel, swings his legs quickly forward between the pommels; and, when they are well over on the other side, he relinquishes his grasp, and lands on the floor in ordinary manner. When practised for the first time, the movement is done so that, after sitting up, the pupil swings his legs slowly between the bars at the command of the teacher. (A similar movement is done with the legs thrown outside the pommels.) Advanced pupils may vault over the bar (or, still better, the horse) in this manner without the saddles. In this case, however, the hands must leave the apparatus almost instantly after they have grasped. They stay only long enough to do the necessary pulling forward. In absence of saddles, the move- ment may be done in ranks with "chain support." For instance, ("*Open ranks, numbers two (one) one step backward — march!*") "*Numbers one (two) chain support — stand!*" The specified numbers place one foot forward, and extend their arms sideways until the hands touch. "*Numbers two (one) — spring!*" These numbers take one step forward, and, placing their hands on the shoulders of those in front of them, they vault between them over their meeting hands, just as the movement is done through saddles. The supports must not lock hands, as some one might catch his feet, and fall, if the hands do not quickly come apart. If benches are at hand, the movement may be arranged as shown in Fig. 212. An exercise which forms an introduction to this form of vaulting is the stoop. fall. 2 F. placing forward (see abdominal exercises).

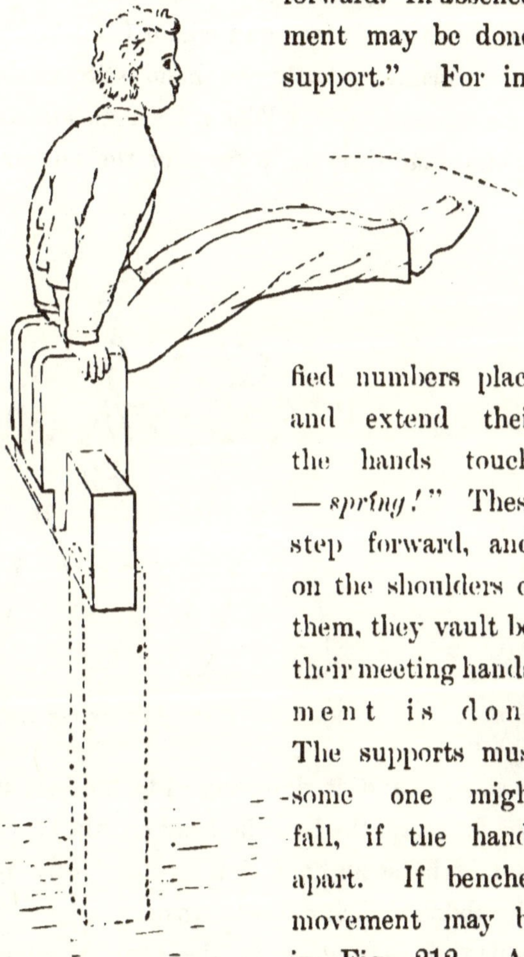

FIG. 211. — VAULTING BETWEEN THE POMMELS.

Vaulting Face Down (Fig. 213). — The bar is put at moderate height (so low that all are easily able to get over). The pupil grasps the bar with both hands, fingers in front, thumbs behind; and by jumping upward, giving a strong pull with his hands, and throwing his legs to one side, he swings himself over the bar, turning his face and whole front toward it. He lands (in usual manner) with one hand still holding the bar and that side turned to it. When his legs are swung to the right, he should land directly in front of the left arm, so that the latter

FIG. 212. — VAULTING BETWEEN THE HANDS.

is straight and at right angles to the bar, and *vice versa*. While the legs pass the bar, the head should be carried high, the arms should be straight, and the whole body form a graceful curve, the abdomen being lower than the feet and head. The hand which is to remain on the bar, in order to do so, has to be turned fingers behind, thumb in front, just before the landing occurs (as shown in Fig. 213). If the hand is already turned in this manner before the start, that arm cannot be so well used for pulling, the hand is apt to slip off the bar — which means accident — and the movement loses its beauty of form. The teacher (or another pupil) should stand on the other side of the bar ready to grasp the one who vaults, and thus to prevent falls. To do this, he stands close to the bar and takes hold of

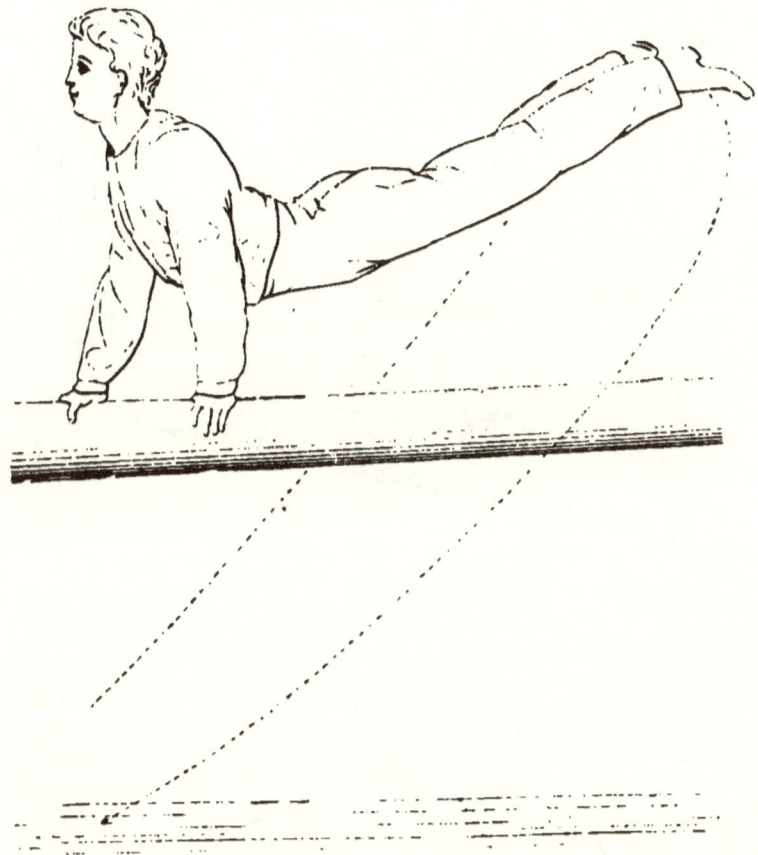

FIG. 213. — VAULTING FACE DOWN: SHOWING HOW THE HAND TURNS JUST BEFORE LANDING OCCURS.

the pupil's wrist with one hand and with the other grasps the same arm from behind and just below the shoulder; he will thus be best able to assist the pupil if need be. As the pupils gain skill, the bar is raised from time to time. The movement is also practised with a short run for start on the same principles as the introd. to vault. ("double start").

Vaulting Face up, or Sit Over (Fig. 214). — This is done on the same principles, except that as soon as the feet leave the

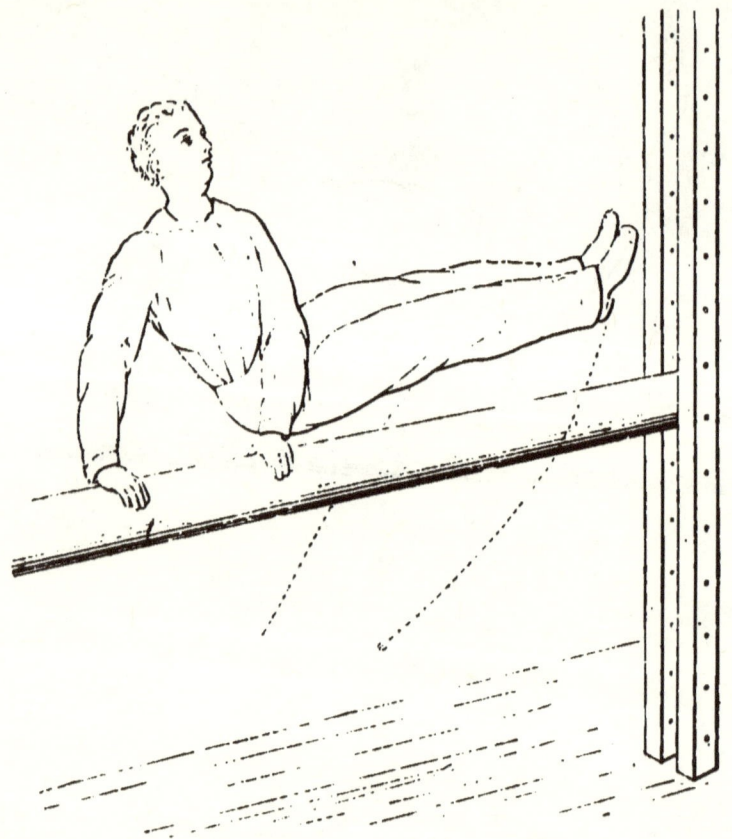

FIG. 214. — VAULTING FACE UP. "SIT OVER."

ground, the face is turned in their direction and the back toward the bar; the position of the body resembles sitting position. When the legs pass the bar, one hand (the left, if the legs are swung to the left, and *vice versa*) leaves it and is quickly swung behind the body, where it grasps the bar close by the other hand (which is now taken off), where it remains until the landing, knee-extension, and lowering of the heels have been done.

The vaulting may also be done so that the body faces the same way throughout the movement. Finally, it may be done with 180° turning; this is executed like vaulting face down, except that the body keeps on turning in the same direction, so that when landing occurs it faces as in "sit over." While the body turns, the hands change places on the bar.

These forms of vaulting may be done as well on the horse or box or other convenient apparatus. On the horse (or box) vaulting may also be done lengthwise (after the manner of "leap-frog"), forward or backward, and in many other ways. To prepare the vaulting lengthwise, it is done halfway, to stride sitt. pos., in the saddle, the pupil sitting off either forward or to the side. Vaulting on the horse is always done with a short run for start.

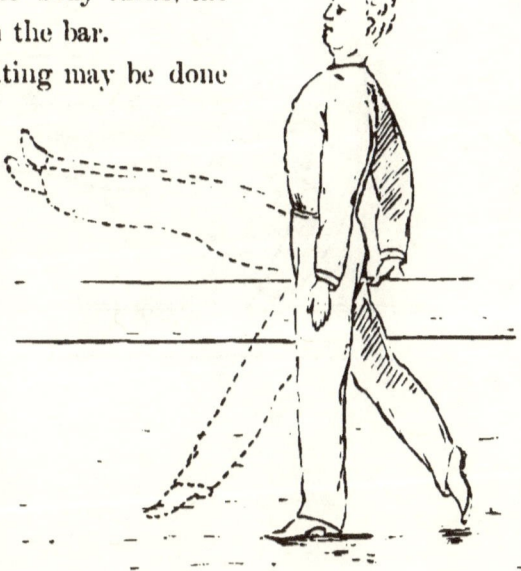

FIG. 215. — SIDEWAYS VAULTING OVER BAR.

Sideways Vaulting over Bar (Fig. 214). — The pupil stands close to the bar with his side turned to it. He grasps the bar with one hand, swings the leg of the same side backward, then quickly forward, and, at the same time, he jumps with the other foot and swings that leg forward too, thus throwing himself over the bar, both feet being brought together as they pass (like "sit over"). Before landing, he changes hands on the bar, so that the one which was outside supports him when he lands.

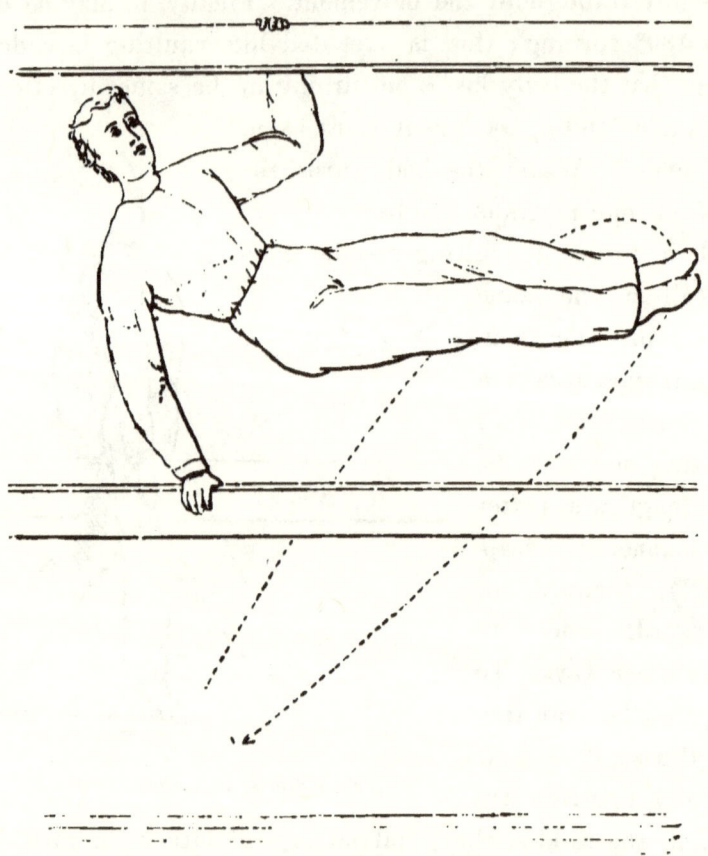

FIG. 216. — VAULTING THROUGH DOUBLE BAR, FACE UP.

The movement may be done so that, instead of on both feet, the pupil lands on the outside foot, and, swinging the other leg forward, he immediately jumps back to the other side of the bar, and continues in this manner from side to side, at the same time moving slightly forward until he reaches the end of the bar.

Vaulting through Double Bar, Face Up (Fig. 216). — The

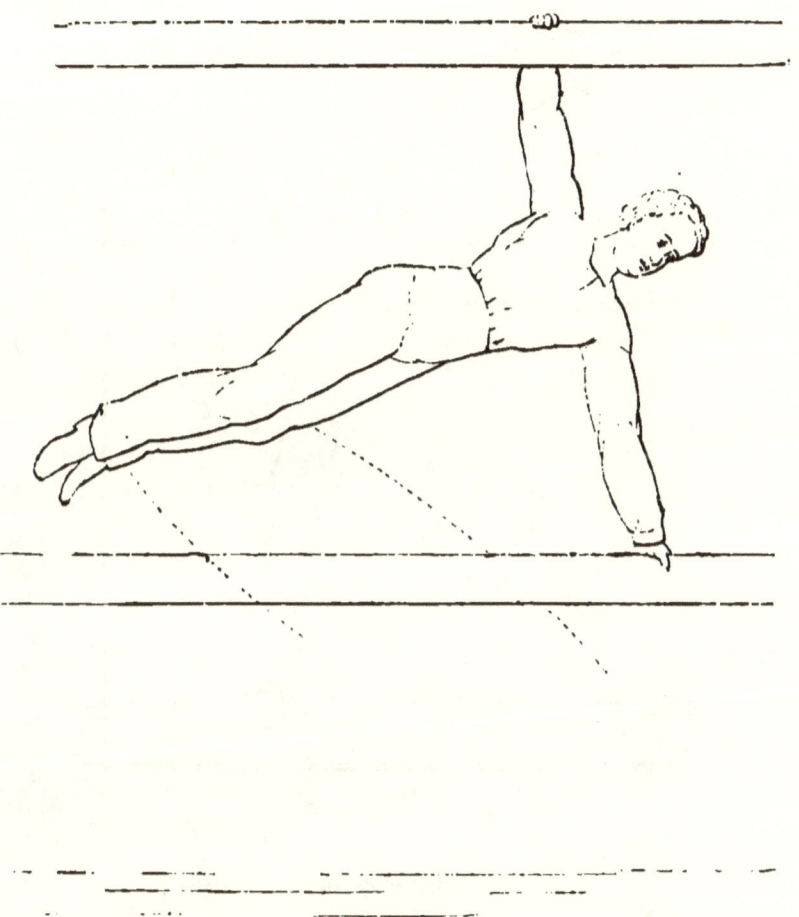

FIG. 217. — VAULTING THROUGH DOUBLE BAR, FACE DOWN.

lower bar is put at usual height, the upper one about four feet higher. The pupil stands with his side turned to the bars and his hand resting on the lower one. Lifting himself on this hand, he now jumps, grasps the upper bar with his other hand, swings his legs forward between the bars, and lands on the other side with one hand on the lower bar. In this way, he

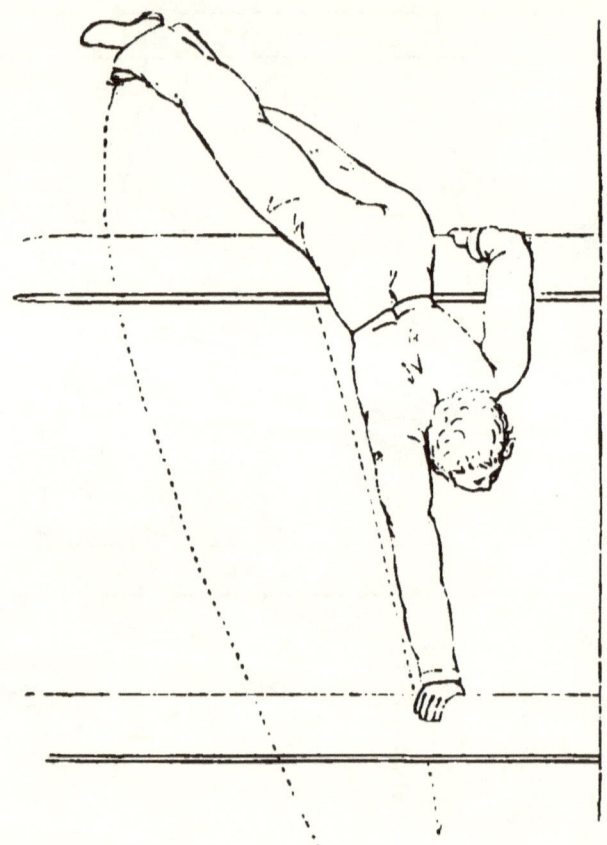

FIG. 218. — VAULTING OVER THE UPPER BAR.

jumps from side to side, using the landing for a new start until he reaches the other end.

Vaulting through Double Bar, Face Down (Fig. 217). — The start is taken with a short run, one hand is placed on each bar, and the vaulting is similar to that over the single bar. The pupil may also face straight forward throughout the movement, or he may "sit over," as previously described.

These otherwise easy movements are made more difficult if the bars are put higher, or if they are placed nearer together.

Vaulting over the Upper Bar (Fig. 218). — The start is taken by running. When the pupil jumps, he places one hand on the upper bar with the fingers forward, the other hand on the lower bar with thumb pointing forward, fingers backward. As soon as the hands have grasped, he throws his whole body to the side of the upper arm, and his legs swing over the upper bar, while he turns the front side of his body to that bar. Pulling with the upper hand, and pushing with the lower one, he tips himself backward over the upper bar, to which he subsequently removes the lower hand, so that he comes into balance hang. pos. (Fig. 207). He then sits off backward (as described in introd. to vault.), and places both hands on the lower bar, when he lands. The bars must at first be put quite close together, the upper one so low that both legs can at once be swung over it (not one at a time as if climbing). The movement is exceedingly difficult.

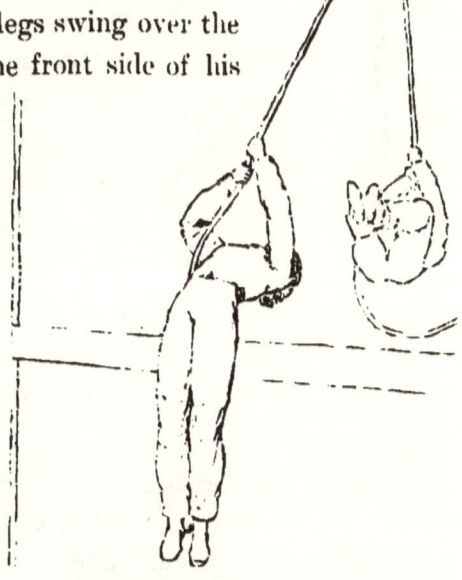

FIG. 219.—VAULTING WITH ROPE.

Vaulting with Rope (Fig. 219). — Like the movements on the double bar, this exercise is a heaving-movement combined with jumping. If a vertical rope is suspended at a convenient distance from the bar, the latter is used to jump over; otherwise the benches (or a horizontal rope) may be arranged as obstacle at a suitable distance (about three steps from where the rope hangs). Let us suppose that the pupil grasps the rope with the right hand. He takes one or two steps backward, stretches the hand that holds the rope as high up as possible, and turns this side toward the obstacle. Now he takes a rapid step forward with the right foot (Fig. 220), and springs from the ground. As soon as the feet leave the ground, the left hand grasps the rope above the right; and, swinging his legs forward, he lets the rope carry him over the hurdle, while he hangs on bent arms. As soon as his legs pass the hurdle, he drops the rope, and lands as in "forward jump." This movement, like all others, should be practised equally to each side. The higher the obstacle, the farther must it be distanced from the rope. If the bar is very high, the start is taken by running, and the hands slide up on the rope when the feet leave the ground.

FIG. 220.—VAULTING WITH ROPE: THE START.

Somersault with Support of the Hands (Fig. 221). — Although this movement is very violent, it may be safely tried by far advanced pupils. When practised for the first time, it is best done on the saddles, the bar being placed near the floor. The pupil places one hand on each pommel, swings his legs backward and up; and, keeping his arms straight, he turns a somer-

MOVEMENTS OF THE SWEDISH SYSTEM 195

sault. Before landing, he pushes well off from the saddle with his hands, and lets go his grasp of the pommels. A pupil

FIG. 221.—SOMERSAULT WITH SUPPORT OF THE HANDS.

should stand in front of each pommel, and give the necessary assistance by grasping the vaulter around the waist and under the shoulder (compare Fig. 222). After some practice, no help is needed, although a pupil should always stand on the other side of the bar, so that accidents may not occur. The movement is also done with a run for start.

Later on the somersault may be executed over the vaulting-box. As a preparatory exercise, the pupil stands on the box, and, placing his hands near the edge, turns a somersault, and lands on the floor, while two other pupils assist him (Fig. 222). As soon as this can be done with ease, the somersault is turned first across (Fig. 223), then lengthwise (Fig. 224) over the box, the start always taken with a short run. At first two pupils give assistance, and stand in a position which enables them to grasp the one who vaults (Fig. 225). Later on, they give no assistance, except at the landing, or when necessary to prevent accident. The somersault may be turned either with utmost speed, or so that the pupil stands balancing on his hands for a moment, and then lets

FIG. 222.—SOMERSAULT FROM BOX WITH ASSISTANCE.

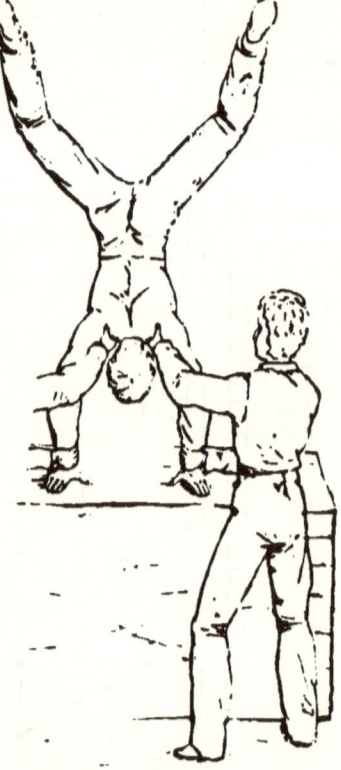

FIG. 223.—SOMERSAULT ACROSS THE BOX WITH ASSISTANCE.

his legs gradually "weigh over." Exercises preparatory to somersaults are deep arch-flexions, reverse stoop fall. pos., etc., etc.

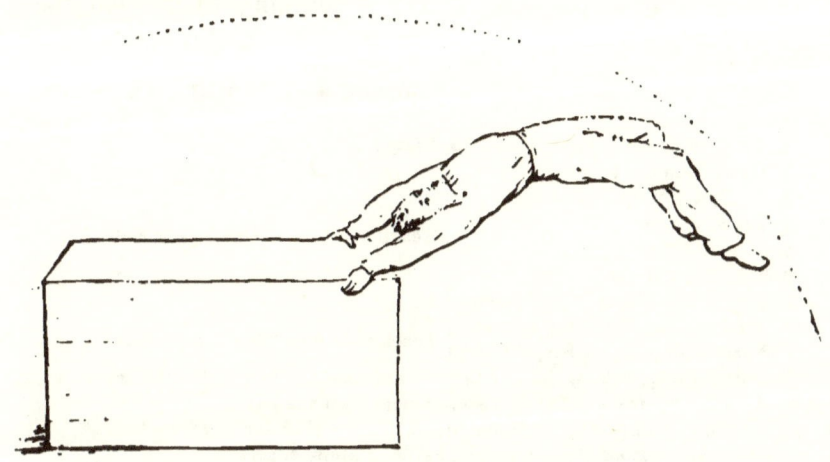

FIG. 224. — SOMERSAULT LENGTHWISE OVER THE BOX.

There are many other forms of vaulting, like pole-vaulting, various movements on the horse, etc., which are used more or less for advanced classes, but which it is not necessary to describe here, the most common forms, as given above, furnishing a sufficient variety to last for many years of practice.

A complete table of exercises should contain both jumping

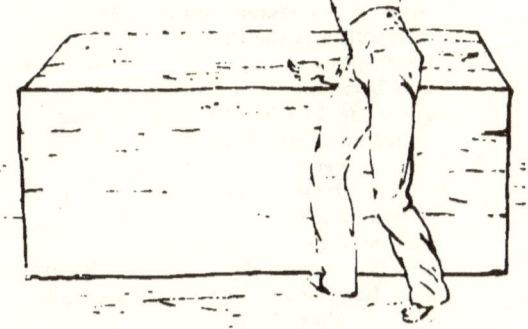

FIG. 225. — HOW TO STAND READY TO ASSIST ONE WHO IS TO TURN SOMERSAULT OVER THE BOX.

and vaulting; the jumping is usually made to come after the vaulting and immediately before the respiratory exercises.

Advanced classes may have several exercises of each kind in every lesson. Yet it should be understood that no vaulting is to be tried until the main forms of jumping have been thoroughly practised.

Progressive arrangement of jumping and vaulting exercises:

1. Prep. to Jump. Teacher counts, "*One !* . . . *Four !*"
2. Upw. Jump.
3. 90° Upw. Jump.
4. Forw. Jump.
5. Sidew. Jump.
6. 2 A. Fling. Upw. Jump.
7. 2 A. and L. Fling. Upw. Jump.
   } Teacher counts, "*One !* . . . *Five !*"
8. Forw. Jump, One Step's Start
9. Forw. Jump, Three Steps' Start
   } Teacher counts, "*One !* . . . *Five !*"
10. Forw. Jump, 1 (2, 3) Step's Start. Command, "*Start!* *Four!* *Five!*"
11. Jump down from Bench. Teacher counts all the parts.
12. 90° Forw. Jump, 1 (2, 3) Step's Start. "*Start!* *Four!* *Five!*"
13. { Jump over Rope (length or height) w. 3 Steps' Start.
    { Prep. to Jump. Command, "*Start!*"
14. { Reach Gr. St. Introd. to Vault.
    { Jump over Rope; teacher tells which foot is to give the final start.
15. { ½ Yd. Gr. St. Introd. to Vault.
    { U$_F$w. Jump. "*Start!* *Five!*"
16. { Prep. to Vault. through Double Bar.
    { Forw. Jump. "*Start!* *Five!*"
17. { Vault. Lengthwise through Double Bar, Face Up.
    { Twice Upw. Jump.
18. { (Introd. to Somersault over the Bar.)
    { Somersault over the Bar.
    { 90° Twice Upw. Jump.
19. { Vault. betw. the Pommels.
    { Jumping fr. Bench, 90° Turn at Landing.
20. { Vault. Double Bar, Face Down.
    { Jumping over Rope with 90° Turn.
21. { Vault. on Horse (box) to Stride Pos.
    { 180° Upw. Jump.
22. { Vault. with Rope.
    { Twice Sideways ("To the Left and Right") Jump.
23. { Vault. Bar, Face Down; later, Face Up.
    { 180° Twice Upw. Jump.
24. { Vault. Lengthwise over Horse.
    { Star Gr. ½ St. Jump. fr. Stall-Bars.
25. { (½ Yd. Gr. St.) Sidew. Vault. over Bar.
    { Jumping Rope, Increased Height.
26. { Vault. Bar, Face Down with 180° Turn.
    { Jumping Rope, Increased Distance.

Repetition of the various forms of vaulting the bar, with a short run for start; the height increasing as well as the distance jumped.
Vaulting over the upper bar.
Vaulting the horse backward and in other complicated ways.
Somersault on saddle.
Somersault on vaulting-box, etc., etc.

RESPIRATORY EXERCISES.

Respiratory exercises are given to lessen the heart-beat, and to make the breathing less labored. They consist of deep respiration, aided by various movements of the arms, in rhythm with the respiratory act, the arm-movement helping to expand the chest at every inhalation, and contracting it at exhalation. By changing the movements of the arms, the chief effect of the exercise can be tolerably well directed to whatever part of the thorax (lungs) may be desired. If any change in rhythm is desirable, inhalation should be done quickly, exhalation slowly; for, whereas quick exhalation may cause dilatation of the heart, quick inhalation is an aid to the heart's action [on account of the negative pressure — less than one atmosphere — on the organs outside the lungs (the heart, the vena cava, etc.)]. As for the much-praised rule of inflating the lungs to their utmost capacity, and holding the breath for a certain length of time before exhaling, it is most pernicious. "It makes your chest grow larger," say its advocates. True, but that enlargement is gained by sacrificing the elasticity of the air-cells, which is a pathological condition known as emphysæma. The method, popular with some, of beating their own chests during deep respiration, or while a large amount of air is confined in the lungs, is equally bad. For, although percussion applied by another person (yet never when the patient *holds* his breath) is most beneficial (the vibration thus caused increasing the contractility of the air-cells), it is at best of no value when applied by one's self; for, to give the blows, you have to fix the muscles which hold the arms to the trunk; and when these muscles are made rigid, while the arms are raised forward, they compress the chest (which interferes with free respiration), and make its walls too rigid to vibrate. As for applying percussion, while you hold your breath after deep inhalation, that is the very best means of distending the air-cells, — the same as you distend an air-filled rubber ball by squeezing it.

To increase the effect of the respiratory exercises, the movement of the arms may be combined with a leg-movement (to accelerate the arterial current into the legs, by the activity of the muscles of the latter), or with a trunk-movement (to increase the chest expansion). Inhaling should always take place through the nose, so that the air may be sifted and free from dust when entering the lungs; whereas exhaling may be done through the mouth, although it is preferable to do that also through the nose.

When an exercise is to be respiratory, the words "*with respiration*" should always precede the command for the arm-movement; and, this being understood, they will not be repeated in the following descriptions.

St. 2 A. Elev. Sidew. — Command, "*Arms sideways — lift!*" The arms are lifted straight sideways into yard pos. (Fig. 82, p. 90). "*Sink!*" The arms are lowered into fundamental position. After the movement is understood by all, it is better to command, "*Arm-elevation sideways — óne! Two!*" . . . Inhale while the arms rise, exhale while they are lowered. The movement raises the whole chest, thus expanding it upward. It may also be done with heel-elevation.

St. 2 A. Elev. Sidew. w. Heel-elev. — Command, "*Arm-eleva-*

tion *sideways with heel-elevation — óne! Two!*" . . . 1. Arms and heels rise. 2. They are again lowered.

Yard St. *d* **2 A. Elev.** — Command, "*Arms sideways — líft! Hands — turn!*" (Fig. 141, p. 128.) *Arm-elevation — one! Two!*" . . . 1. The arms are raised into str. pos. 2. They resume yd. *d* pos. Inhale while the arms rise, exhale while they are lowered. The movement expands the lateral parts of the chest, while it slightly contracts the superior anterior (clavicular) region. The greatest expansion being at the base of the lungs, a sort of "abdominal breathing" occurs. The movement is always done with even rhythm.

On the same principles are executed:

Yard *d* { Turn st. / Turn stride st. / Turn fallout st. *a* / Fallout st. *b* } 2 A. elev.

In the fallout st. positions, the trunk should be erect: hence, command, "Left (r.) foot, large step, backward — pláce!" etc.

Yard *d* **St. 2 A. Elev. w. Heel-elev.** is done on the same principles as 2 A. elev. sidew. w. heel-elev.

Yard *d* **Toe St. 2 A. Elev. w. 2 Kn. Flex.** — Command for commencing position, "*Arms sideways — líft! Hands — tûrn! Heels — líft!*" and for the movement, "*Arm-elevation with knee-flexion — one! Two!*" . . . 1. The arms rise into str. pos., while the knees bend to right angles (inhalation). 2. Commencing (yd. *d* toe st.) position is resumed (exhalation).

St. 2 A. Elev. Sidew. Upw. — Command, "*Arm-elevation sideways upward — one! Two!*" . . . 1. The arms are raised through yard pos. into str. pos. 2. They are lowered sideways downward, and carried as far backward as possible. The turning of the hands is done gradually both when the arms are raised, and when they are lowered.

St. 2 A. Elev. Sidew. Upw. w. Heel-elev. — The heel-elevation begins just before the rising arms reach str. pos., and the lowering of the heels as soon as the sinking arms have passed yd. pos. Command, etc., on the same principles as previous movements.

St. 2 A. Elev. Sidew. Upw. w. 2 Kn. Flex. — Command, "*Arm-elevation sideways upward with knee-flexion — one! . . . Four!*" 1. The arms are raised into yd. pos., while the heels are lifted. 2. The arms rise into str. pos., while the knees bend to right angles. 3. The arms are lowered into yd. d pos., while the knees are straightened. 4. The arms sink into fund. pos., while the heels are lowered. After a day or two of practice, the movement is done in two counts instead of four. This is the true form of the movement; and inhalation is done while the arms rise ("*One!*"), exhalation while they sink ("*Two!*").

St. 2 A. Elev. Forw. Upw. (Fig. 226). — Command, "*Arms forward upward — lift!*" The arms are lifted straight forward upward through reach pos. into str. pos. "*Sideways downward — sink!*" The arms are lowered through yd. pos. into fund. pos. When the arms are raised forward upward, the chest becomes vaulted forward and raised upward; it becomes no wider, but its capacity grows larger in a vertical direction. The following movements are done on the same principles:

Turn fallout st. *a*
Fallout st. *b* } 2 A. elev. forw. upw.
Toe sup. fallout st. *b*

The arm-movement may also be combined with heel-elevation:

St. 2 A. Elev. Forw. Upw. w. Heel-elev., the heels being raised just before the arms get into str. pos., and lowered just before the arms reach fund. pos. Command and execution are, in other respects, similar to the movement just described. The arm-movement may also be combined with a trunk-rotation:

St. 2 A. Elev. Forw. Upw. w. T. Rot. — Command, "*Arm-elevation forward upward, with trunk-rotation to the left (r.) and right (l.) — one! . . . Four!*" 1. While the trunk is rotated to the left (r.), the arms are raised (inhalation). 2. While the trunk is rotated forward, the arms are lowered (exhalation). 3. As No. 1 to the opposite side. 4. As No. 2. The movement may be done from **st., close st., or stride st. pos.** In rotated position, the chest is more expanded than when facing forward; hence this movement is stronger than the corresponding one without rotation.

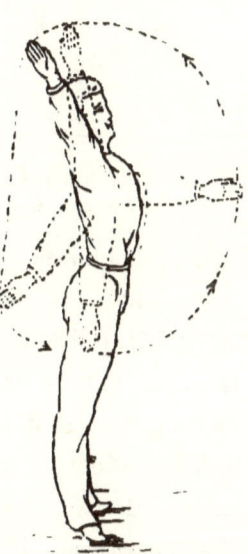

FIG. 226. — ST. 2 A. ELEV. FORW. UPW.

St. 2 A. Ext. Sidew. — The usual 2 A. ext. may be made into a respiratory exercise, if inhalation is done when the arms extend, exhalation when they bend, the arm-movement following the rhythm of deep respiration. But to get the most chest-expansion out of this movement, the palms of the hands should be turned up as in yd. d pos. (Fig. 134, p. 123), for in this position the tendon of the *pectoralis major* muscle is pulled backward and partly wound around the *humerus;* the tension thus caused in the muscle elevates the sternum and pulls the upper ribs apart, thus expanding the chest. Hence, a better respiratory exercise is the

St. 2 A. Ext. to Yd. d Pos. (Fig. 134, p. 123). — Command, "*With palms up, arm-extension sideways — one! Two!*" . . .

The movement is done slowly; and the flexion is not as rigid nor as complete as in the rapid 2 A. ext., the elbows not being brought close to the waist.

Str. Arch Sup. Stride St. 2 A. Ext. Upw. (Fig. 227). — The bar is placed below hip-height, and (to the teacher's command) the pupil takes str. stride st. pos., and bends backward to horizontal position over the bar, so that it goes across the middle of his back, allowing him to rest there. When the bar is low the pupil has to bend his knees. "*Arm-extension upward slowly — one! Two!*" ... The arms bend only to yard e pos. (Fig. 137, p. 125). Inhalation is done when the arms extend upward. In the backward flexion, the chest becomes forcibly vaulted, and the effect of the arm-extension is thus increased. The position would accelerate the heart-beat, if the trunk were not supported by the bar.

FIG. 227. — STR. ARCH SUP. STRIDE ST. 2 A. EXT. UPW.

Yard St. 2 A. Fling. (Fig. 135, p. 124). — Command, "*Arms forward bend and sideways fling — one! Two!*" ... Inhalation takes place while the arms are quickly extended, exhalation while they are bent with moderate speed. This movement expands the clavicular region of the chest (broadens it), a portion usually abnormally flat in persons not possessing physical culture. The apex of the

FIG. 228. — YARD a TURN FALLOUT ST. a POS.

lungs becomes enlarged and invigorated. The movement is used a great deal and for a long time for beginners, since it develops just that part of the chest where they are the most hollow. On the same principles, the following are done:

Yard *a* { Turn fallout st. *a* (Fig. 228); Fallout st. *b* (Fig. 229); Turn st. or stride st. } 2 A. fling.

Yard Walking 2 A. Fling. — Command, "*With arm forward flexion and sideways flinging, forward march — one! Two! Three!*" ...

1. Left foot moves a step forward, while the arms are placed in yard *a* pos. 2. The arms remain in this position while the right foot takes a step forward. 3. The left foot moves forward while the arms extend, and so on as far as the size of the room will allow. The marching is necessarily slow, but yet determined and in good posture. The effect of the arm-flinging, of restoring free respiration, is thus combined with that of slow marching, — mechanical propulsion of the blood into the legs, without increasing the heart's action.

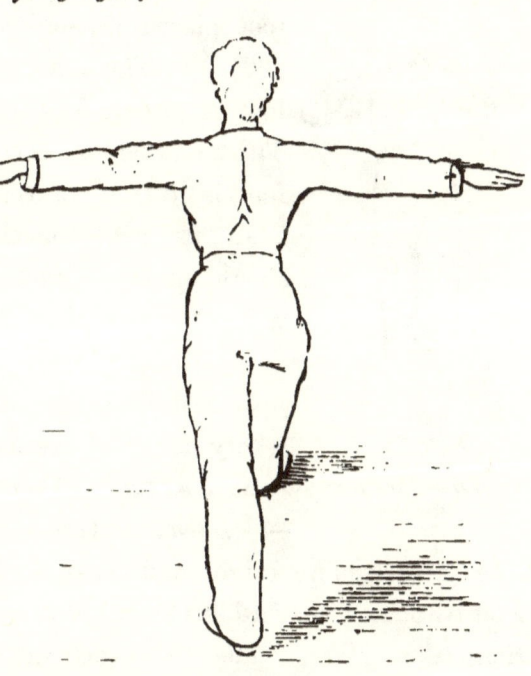

FIG. 229. — YARD *c* FALLOUT ST. *b* POS.

Yard Arch St. 2 A. Fling. (Fig. 230). — Command, "*Arms* *forward — bénd! Slightly, trunk backward — bend!*" The trunk (and head) is bent a little backward. "*Arm-flinging — óne! Two!*" . . . The trunk being bent backward, the chest is more expanded than in the st. fund. pos. The movement may be combined with a leg-movement:

Yard Arch Walk St. *b* **2 A. Fling. w. Ch. of F.** — Command, "*Arms sideways lift and left (r.) foot forward — pláce! Trunk, slightly, backward — bend! Arm-flinging with change of feet — one! Twó!*" . . . 1. The arms are bent, and the forward foot placed beside the backward one (Fig. 230). 2. The arms extend, while the other foot is placed forward into walk st. *b* pos. The trunk-flexion becomes deeper, when one foot is in front of the other, than when the heels are held together. The 2 A. fling. should first be practised in walk st. *b* pos. with change of feet, then in arch walk st. *b* pos. without change of feet, and finally as just described. In its complete form it is a very powerful exercise.

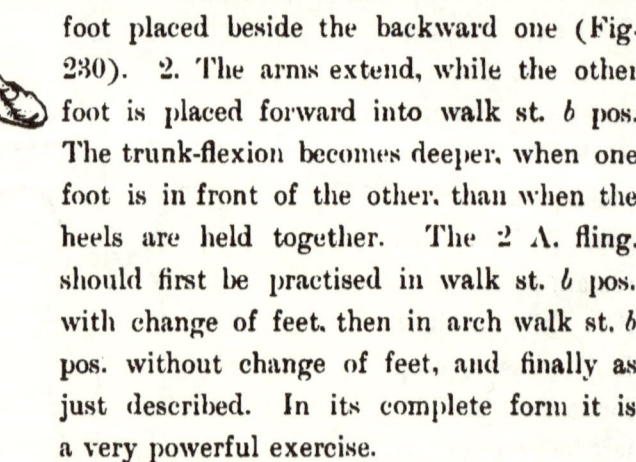

FIG. 230. — YARD a ARCH ST. POS.

St. 2 A. Circ. — Command, "*Arms sideways — strétch! Arm circumduction — start!*" The arms remain straight, and move so that the hands describe a small circuit upward backward, downward forward. Inhalation takes place, while the arms move upward backward, exhalation while downward forward, the movement of the arms following the rhythm of slow respiration. The movement should be communicated through the whole chest as much as possible, the shoulders rising with the arms, etc. Care should be taken not to move the arms too far forward. The movement is continued until the teacher commands, "*Stóp!*"

On the same principles are executed:

Yard c { Arch st.
Turn st.
Turn stride st. } 2 A. circ.

For advanced classes, when the jumping and vaulting exercises become more violent or numerous, it is often necessary to give two respiratory exercises at the end of the lesson. These are then made as different as possible; for instance, one is a 2 A. fling., the other a 2 A. elev., the commencing position of each differing from that of the other. Moreover, a complete table of exercises usually begins with an easy respiratory exercise, which has been done on some previous occasion. For, although the oxygen inhaled is not turned to muscular work in the same lesson, yet it is a good plan to furnish a storage of it for future use, since a lesson in gymnastics will exhaust much of the supply already present. Furthermore, a respiratory exercise may be applied any time during a lesson, when the teacher deems it necessary.

Progressive arrangement of respiratory exercises:

1. St. 2 A. Ext. Sidew.
2. Yd. a St. 2 A. Fling.
3. St. 2 A. Elev. Sidew.
4. St. 2 A. Ext. to Yd. d Pos.
5. Yd. d St. 2 A. Elev.
6. St. 2 A. Elev. Forw. Upw.
7. St. 2 A. Circ.
8. St. 2 A. Elev. Sidew. Upw.
9. Yd. Walk St. b 2 A. Fling. w. Ch. of F.
10. St. 2 A. Elev. Sidew. w. Heel-elev.
11. Yd. d St. 2 A. Elev. w. Heel-elev.
12. Yd. Fallout St. b 2 A. Fling.
13. St. 2 A. Elev. Forw. Upw. w. Heel-elev.
14. St. 2 A. Elev. Sidew. Upw. w. Heel-elev.
15. Yd. d Toe St. 2 A. Elev. w. 2 Kn. Flex.
16. Yd. Walking 2 A. Fling.
17. Fallout St. b 2 A. Elev. Forw. Upw.
18. Yd. Turn St. 2 A. Fling.
19. Yd. d Fallout St. b 2 A. Elev.
20. Yd. Turn St. 2 A. Circ.
21. Close St. 2 A. Elev. w. T. Rot.
22. Yd. Turn Stride St. 2 A. Fling.
23. Yd. d Turn St. 2 A. Elev.
24. Yd. Turn Fallout St. a 2 A. Fling.
25. Yd. Turn Stride St. 2 A. Circ.
26. Toe Sup. Fallout St. b 2 A. Elev. Forw. Upw.
27. Yd. d Turn Stride St. 2 A. Elev.
28. Yd. Arch St. 2 A. Fling.
29. Stride St. 2 A. Elev. w. T. Rot.
30. Yd. d Turn Fallout St. a 2 A. Elev.
31. Yd. Arch St. 2 A. Circ.
32. Yd. Arch Walk St. b 2 A. Fling w. Ch. of F.
33. Str. Arch Sup. Stride St. 2 A. Ext. Upw.

It should be remembered that the respiratory exercises are of so varied a nature that an absolute progression can hardly be made, since it is difficult, if at all possible, to compare two movements of different character.

RULES FOR MAKING TABLES OF EXERCISES.

One of the most difficult duties of a teacher of gymnastics is that of making out tables of exercises — and every teacher is expected to make his own. He should proceed in a manner somewhat after the following:

First cut a number of papers ("cards"), each large enough to hold the names of from nine to fifteen exercises written under each other. Number each card. Each slip of paper is intended to hold one table of exercises; and twenty-four of them will suffice; for, counting the winter season (during which the exercises are usually taken) as six or seven months, twenty-four tables of exercises will be enough to last throughout the season, since one table is to be used for a week at a time (besides, legal holidays and other intermissions must be counted out). Next estimate the nominal average of physical development of the class. To do this, there must be taken into consideration the individualities of age, sex, strength, nationality, etc., etc. After making this, fill in the exercises on the "cards" on the following plan: for instance, take the shoulder-blade movements. Look at the progressive table of these exercises, as given in this book,[1] and estimate about how far the class in question will be able to proceed in these exercises. Let us suppose exercise No. 15 to be considered the limit of the progress in one year; write this exercise down in the proper place on card No. 24.

[1] These progressive tables for each class of movements form one of the features of this hand-book which make it differ from all previous works on Swedish Gymnastics, not only in English, but in other languages as well.

and then spread the other fourteen exercises over the remaining twenty-three cards, letting the difficult movements appear twice or more. In the same manner, treat all the other classes of exercises, taking into consideration what apparatus there is at hand, what time is allotted to each lesson, how many lessons a week there are to be, etc., etc. In this way, the teacher "constructs" a set of tables suitable for the class in question; and, in the course of his work, he should try to follow the tables as closely as possible: yet he must be prepared to make whatever alterations circumstances render necessary. For example, he may find it expedient to keep one exercise through a greater number of tables than he had at first planned; or he may have to postpone a movement on account of its producing too many faults, etc., etc. For, a set of tables that may have suited one class of pupils may not always suit another, even though the scholars be of about the same age and average development; nor can one teacher always accomplish exactly what another would under the same conditions, even if he uses precisely the same tables of exercises; for there is just as much individuality in teachers as in pupils — where one has succeeded another will fail, and *vice versa*. For this reason, the tables printed at the end of this book must be considered only as samples, and not as regulation tables to be mechanically followed. An inexperienced teacher may begin by using these tables as they are written, but he will soon be competent to make his own and thus to put his individuality into the work, greatly to his advantage. For, a teacher of gymnastics will be just as unsuccessful as any other teacher, if he makes himself into a machine.

If apparatus is at hand, each table should contain: (1) Introductions. (2) Arch-flexions. (3) Heaving-movements. (4) Balance-movements. (5) Shoulder-blade movements. (6) Abdominal exercises. (7) Lateral trunk-movements. (8) Slow leg-movements. (9) Jumping and vaulting. (10) Respiratory exercises, in the order enumerated. The first few tables, which for beginners must be very simple, naturally cannot contain all these classes; thus it will not be until the seventh or eighth table that the order can be strictly followed; and, on rare occasions, deviations from the rule may be made according to the teacher's judgment.

As the work progresses, various additions may be made to the tables. Thus, for introductions, may be used resp. exers., Leg-movs., A.-exts., lateral T.-movs., and exercises for attention and rhythm. one or more of each kind in every table, these being chosen from exercises which have been done before. Moreover, another heaving-movement may be added just after the slow leg-movements; several jumping and vaulting exercises may be placed in the same table; two or more lateral trunk-movements (as different as possible) may be used, etc., etc.; so that a lesson for an advanced class may contain as many as twenty different exercises, whereas one for beginners should have only six or seven.

If no apparatus is at hand, some kinds of exercises (heaving-movements, vaulting, etc.) have to be entirely omitted, and others will have to be used as substitutes. This somewhat changes the order of exercises in each table, and two or more of the same class have to be applied in the same lesson. These should then be placed as far apart, and made as different in character, as possible. Under such circumstances, the free-standing arch-flexions may be placed in the middle of the table, and on each side of them leg-movements, lateral trunk-movements, shoulder-blade movements, etc. In a schoolroom where for some reason the teacher is not allowed to use the chairs and desks as apparatus, the tables have to be made out according to the plan just described.

If the exercises are to be taken between the recitations for a period of about five minutes, more or less, there will not be time for more than a leg-movement, a 2 A. ext., an arch-flexion, and a lateral trunk-movement, or four or five other exercises chosen according to the needs of the class. This arrangement may do for very small children (six or seven years of age), but for all others it is of small value, even though it be better than no exercise at all. To have any lasting effects, the gymnastic exercises should occupy at least half an hour a day. And it would be strange if that much time could not be easily spared from the theoretical exercises for the sake of the physical welfare of the school-children. The fact is, that this "loss" of time would be an everlasting gain, inasmuch as the mind is capable of more and of better work, if it dwells in a healthy — a well-balanced — body. In the history of man, the fact stands out in bold relief, that a race, mentally gifted or physically developed, has survived, only if its representatives were physically stronger than their competitors; whereas, a physically degenerated race, no matter how brilliantly endowed otherwise, soon ceases to hold the lead in culture and civilization. The histories of the ancient peoples all bear witness to this.

Marching and running should be applied in every lesson, if the room admits of such exercises. They are best put in the middle of the table, just after the shoulder-blade movements (or, if two shoulder-blade movements are used, between the two); and they should be immediately followed by a slow leg-movement or a respiratory exercise, whenever the teacher deems it necessary. If possible, the marching and running should always take place out-of-doors (in the school-yard, in the yard connected with the gymnasium, etc.), partly because there is more space, but especially because there is more air, more oxygen.

When tables are made out for advanced pupils, — those who have practised gymnastics for several years. — it is not safe to begin at once with heavy exercises, even though they have been done before. If there has been a month or more of rest before the lessons begin, the first few tables should have an introductory character; but the progression may instead be made more rapid. Each step forward in the various classes of exercises is made longer; and every exercise, previously practised and now repeated, should be done to utmost perfection. In this way each movement may, from year to year, in a measure serve as its own progression.

It is desirable that the teacher, before beginning the lesson, should make himself so familiar with the table of exercises he is going to use, that he knows it by heart. A teacher who has many different classes, however, could not well commit to memory all the tables: for that reason he may carry a memorandum of tables written on small cards (the size of a visiting-card), which he can easily hold in his open hand, and yet use that hand to correct or to illustrate a movement. He should take care, however, not to consult these cards too often, as he is expected to have his eyes on the class almost incessantly, so that attention and discipline may not become slackened.

PART III

APPENDIX

APPENDIX

MEDICO-GYMNASTIC EXERCISES USED IN EDUCATIONAL GYMNASTICS.

In every class pupils will be found who, through deformity or unusually poor development, are unable to proceed as rapidly as the others. To enable these to keep with the class, and to prevent them from checking the progress of the others, — since none are to be excused from gymnastics, except on account of incurable deformity or organic disease, which would be aggravated by exercise, — the teacher resorts to movements of a stronger and more corrective character, which he can apply individually to those especially needing them. These exercises are borrowed from medical gymnastics, and hence are to be considered as an appendix to educational gymnastics. Only the simplest and most common forms will be described, since it requires a special education to intelligently apply medical gymnastics — a knowledge that cannot be obtained by merely reading about the subject, and which cannot be expected in the average teacher of gymnastics.

Round Shoulders. — When the shoulder-blade movements fail to accomplish their object, the teacher applies the following exercise: The pupil stands facing the teacher with his arms in yard pos., the palms of the hands turned forward. The teacher (in walk st. *b* pos.), standing close to the pupil, grasps the latter's arms near the hands (over grasp, if the pupil is small; under grasp, if he is tall), and pulls them forward into reach pos. under a moderate resistance. The pupil moves his arms back into yard pos. under resistance from the teacher; and, when the limit of voluntary motion is reached, the teacher presses the arms just a little farther backward. The movement, which is repeated five or six times, should be done evenly and not by little starts and pushes; and the teacher should see that the pupil maintains good posture all through it. The exercise is called **yd. st. resistive A. add. and abd.** (Fig. 231). As the

FIG. 231.— YARD ST. 2 A. ADD. AND ABD.

pupil gains strength, the commencing position should be made more difficult, forw. lying, stoop st., fallout st. *b*, and foot gr. fallout st. *b* (Fig. 232) positions being used in the order enumerated. If the pupil cannot well maintain the standing position, he may sit while doing the movement; and, instead of stoop st. pos., the corresponding sitting position (Fig. 233) may be used.

Drooping Head. — If the free-standing head-flexions, etc., are not sufficient to make the head stay in good erect posture, the following movement is applied: The pupil stands facing the teacher, who places his hands, one upon the other, behind and against the pupil's head (Fig. 234), letting his forearms rest on the anterior part of the pupil's shoulders. The pupil bends his head forward and then backward as far as it will go, the teacher offering a moderate resistance against the backward flexion, thus bringing the muscles of the neck into stronger activity. When the head bends backward, the chin should be

APPENDIX

FIG. 232.— Yard Foot Gr. Fallout St. b 2 A. Add. and Abd.

FIG. 233.— Stoop Sitt. A. Abd. and Add.

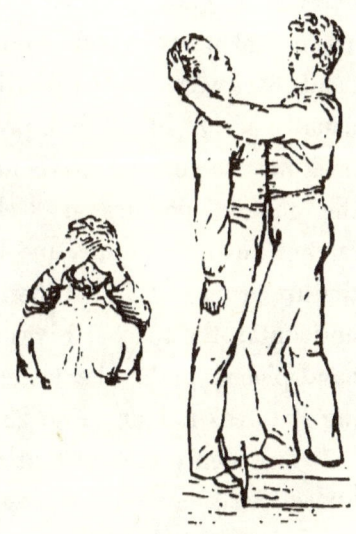

FIG. 234.— St. Neck Flex.

drawn in, for then the insertion (gymnastically speaking) or origin (anatomically speaking) of the sterno-cleido-mastoid muscle is moved backward, and a tension is produced in this muscle which causes an elevation of the superior part of the sternum. Thus the exercise not only corrects the posture of the head, but it also tends to expand the chest. The movement is repeated five or six times in succession. The reach gr. st. pos. may also be used as commencing pos.; the teacher then stands on one side of the pupil, steadies him with one hand, and resists the neck-flexion with the other. The movement may also be taken in a stoop fall pos., obtained from the reach gr. st. pos. by moving the hands a few bars downward and the feet backward on the floor (or on a bench or other elevation), the position growing in strength as it comes nearer to horizontal stoop fall. pos. When done in stoop fall. pos., the exercise has a far more extended effect; for now it is also an abdominal exercise, a good posture calling for strong contraction of the abdominal walls, which contraction must increase as the head bends backward, or the position is lost. As a drooping head is usually accompanied by a protruding abdomen, the stoop falling neck-flex. is very effective in correcting not only the head but the general posture as well. Besides, this exercise reaches farther down the spine, and is more efficacious in straightening the upper part of it. When the stoop falling position is nearly horizontal, the teacher gives the pupil a slight support by one hand placed under the latter's abdomen, the other hand resisting the neck-flexion (Fig. 235).

It has been proposed that this exercise be done by the pupil himself making resistance against his own movement, with his hands locked behind his head. The effect of such a method is to strengthen some of the muscles of the arm; but for invigorating the muscles of the neck, it is absolutely worthless. The whole procedure forcibly reminds us of how the famous Baron Münchausen pulled himself and horse out of a swamp by taking hold of his own "pigtail," and by that means lifting himself and horse until they reached firm land.

Flat Chest. — This is best corrected by heaving-movements

FIG. 235. — STOOP FALL. NECK-FLEX.

and respiratory exercises; if, however, it is desirable to hasten the effect, the following exercises can be used:

1. **Wg. St. Heel-elev. w. Chest Expansion.** — The pupil takes "hips firm." Standing behind him, the teacher takes hold of his elbows; and, while the pupil rises on tip-toe and inhales, the teacher pulls his elbows backward as far as possible without tipping him backward. In this way the chest becomes forcibly (and passively) expanded, so that it can hold more of the

in-rushing air. The pupil exhales, and lowers his heels, while the teacher lets the elbows resume commencing position. The movement is repeated from five to ten times. It is best to use the command, "*Heel-elevation with respiration — óne! Two!*" . . .

2. **Str. Gr. St. Chest Expansion** (Fig. 236). — The pupil stands close to the stall-bars, with his back turned to them, and grasps as high as he can reach without rising on tip-toe. The teacher stands in front of him, and, bracing himself with one foot against the bars, he applies both hands to the pupil's back (between the shoulders). By letting his hands slide downward and forward, he pulls the pupil's trunk forward into arch pos., and in so doing lifts his heels from the floor. During this operation, the pupil inhales; and he exhales as the teacher lets him fall gently back into commencing position. The movement extends the whole front of the body, and vaults the chest forward. If vertical poles (whose lower ends are fixed) are at hand, the pupil takes str. gr. st. pos. between them; and, standing behind, the teacher pushes the pupil's thorax forward and upward with one hand applied in the middle of the back just under the shoulder-blades (Fig. 237). When done in this manner, the movement gives a little more elevation to the chest than is attained when the stall-bars are used. A narrow doorway will answer the purpose of vertical poles.

Fig. 236. — Str. Gr. St. Chest Expansion.

3. **Reach Lying Chest Expansion** (Fig. 238). — The pupil lies face up on a bench (or other elevation) and places his arms

in reach pos. The teacher, standing behind, grasps the arms around the wrists and pulls them backward into str. pos. while the pupil inhales, and pushes them forward into reach pos. while he exhales. The movement — which is repeated six or eight times — expands the lateral parts of the chest and vaults it upward. The effect is increased, if the pupil makes a slight resistance while his arms are being pulled backward.

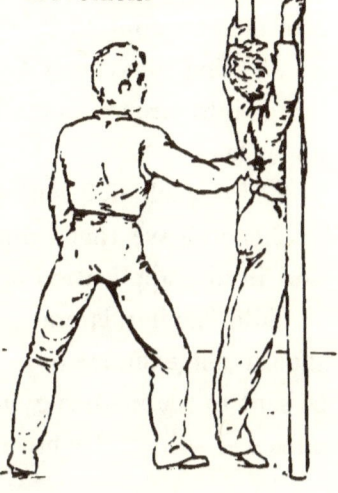

FIG. 237. — STR. GR. ST. CHEST EXPANSION.

Lordosis. — By this name we understand that condition of the spine where the lumbar region is too much convexed (forward) and the pelvis tipped forward in proportion. It is caused by weakness of the muscles of the abdomen and upper front leg, or (although rarely) by weakness of the extensors of the back. The chief aim of specialized exercise should be to strengthen the muscles of the abdomen and upper leg. If this is not achieved by the ordinary abdominal exercises (and movements in crook hang. and cr. ½ st. positions), the following movement may be applied:

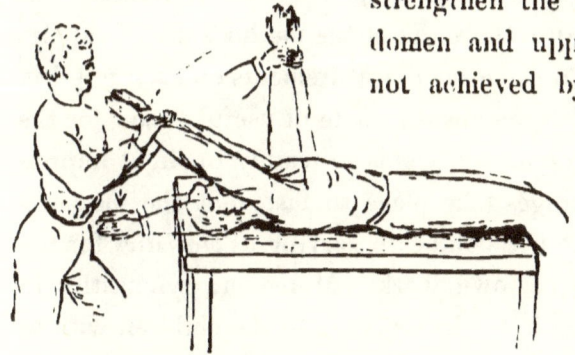

FIG. 238. — REACH LYING CHEST EXPANSION.

Str. Gr. Ly. 2 L. Elev. w. Resist. — The pupil takes the stretch grasp lying position, and raises his legs as high as he can (Fig. 156, p. 139), while the teacher makes a slight resistance against this movement by grasping the rising legs around the ankles with one hand and with the other exerting a gentle pressure on the abdomen. The teacher presses the legs back into commencing position, while the pupil makes a slight resistance. The movement (which is repeated three or four times) brings into strong contraction the muscles of the abdomen and upper leg.

As for such deformities as lateral curvature, we advise the teacher to leave them alone unless he has had special instruction in the application of "Swedish movements" to such cases. A child having lateral curvature or other serious deformities should not exercise in a class with healthy children, but should be sent to the medico-gymnastic specialist for treatment; thus, a teacher will hardly have occasion to do anything in such cases.

RULES FOR MEASURING PUPILS.

One of the features of a modern gymnasium is the measuring of the pupils at the beginning and at the end of every working year, so as to observe what development each individual has gained — *i.e.*, to measure the value of the methods used in applying the exercises. To make such measurements on each pupil in a large school would be an absurd waste of useful time; for the human body during childhood yields so readily to slight impressions, and radical changes take place so fast, that the measurements can be of but little value in serving as *statistics* for the teacher to judge of his own work. Moreover, gymnastics in schools must be essentially class work, and the children will, as a rule, be grouped together according to age or size; and individualization can be done only to a very slight extent; hence the teacher could have but little use for the measurements towards supplying individual exercises. In gymnasiums where individual work is a distinctive feature — gymnasiums for grown persons — the conditions are a little different; for here the teacher is expected to give each pupil a separate set of exercises suited to his particular development (or lack of it), and to do so the teacher must measure each individual and make his prescriptions accordingly, the figures received at the beginning of the course telling what should be done, those at the end what has been done. Thus it will be seen that the figures are still for the teacher, — not . for the pupil, — although he may show them to the pupil, if he so desires.

The records of these measurements taken on grown persons should be kept by the teacher, so that, with these as a standard he can make the necessary improvements in the system of gymnastics which he uses. Moreover, they will furnish coming scientists with the necessary statistics for their study of the evolution of man, etc.

Since Man is himself the unit by which his development should be measured, some measure — usually the height — should be taken as a unit, and all the others compared to this and recorded in fractions (percentage) thereof. This is the only way in which we can be able to judge concerning the harmony of the development; for we may now compare these figures to a standard table of measurements prepared by taking the average of a great number of measurements made on well-proportioned bodies. For ease of calculation and in conformity with progressive ideas, all measurements should be taken according to the metric system.

To measure every part of the body would be a waste of time; for, if a few of the important parts are measured, the figures will give a fair estimate of the general development. Parts to be measured are:

1. *The height of the whole body.*

2. *The width of the chest*, taken on a level with the nipples, just in front of and close to the arms when they are hanging straight down.

3. *The depth of the chest*, taken on a level with the nipples, the arms of the caliper being on a level.

4. *The circumference of the chest*, also taken on a level with the nipples. Two measures are taken: one with the chest inflated, the other after exhalation. The tape-measure should be placed horizontally around the chest.

5. *The shoulder width*, taken outside the acromion, so that the caliper touches this bone.

6. *The waist.* — The circumference is measured in the narrowest place.

7. *The width of the hips.* — The pupil is in close st. pos. while the measure is taken between the most protruding points of the hips — the greater trochanters of the femurs.

8. *The inside length of the leg* from the perineum to the ground, the feet being slightly apart and the body carried by both legs equally. On women, the outside length of the leg from the greater trochanter to the ground is taken instead. If the legs are of unequal length, a note is made of this.

9. *The length of each arm* from the acromion to the tip of the middle finger.

Furthermore, the pupil's weight and age are recorded.

The record charts may conveniently be given the following appearance:

Name.	Age.		Weight.		Height.		The Chest.									Remarks.
							Width.	Depth.	Circumference. Inflated.	After Exhalation.	Width of Shoulders.	Waist Circumference.	Width of Hips.	Inside/Outside length of Leg.	Outside length of Arms.	
	Years.													Right.	Right.	
	Months.													Left.	Left.	
		Month. First														
		Last														

The following chart, extracted from a similar one published in a text-book on Anatomy by Prof. T. Hartelius (Stockholm, 1884), may serve as a standard for comparison. The figures represent an average of numerous measurements taken by Prof. Carl Curman on living models and on antique statuary.

AVERAGE PHYSICAL DEVELOPMENT, WHEN THE BODY IS COMPARED TO ITSELF.	HEIGHT = 100.			
	Man.	Woman	Child.	
	25 years old.		About 5 years old.	About 1 year old.
Height of the head above the chin (measured from the top of the skull to under the chin)	13.6	14.0	19.0	22.5
Height of the head above the top of the sternum (from the top of the skull)	17.5	17.6	21.5	25.5
Height of the head above the acromion	17.7	18.2	21.0	25.0
Height of the head above the inferior end of the sternum,	27.5	28.5	31.5	36.5
Height of the head above the crest of the ilium	41.5	42.5	47.5	53.0
Height of the acromion above the greater trochanter	30.0	30.0	32.0	33.5
Length of the arm from acromion to tip of middle finger	45.5	44.0	42.0	41.5
Length of the arm from acromion to the middle of the elbow	20.0	19.0	18.0	18.0
Length of the forearm from the elbow to the middle of the wrist	14.5	14.0	12.5	12.0
Length of the leg from the ground to the crest of the ilium	58.5	58.0	53.0	46.5
Length of the leg from the ground to the perineum	47.5	–	40.0	35.0
Length of the leg from the ground to the middle of the patella	28.5	28.0	26.5	24.0
Height of the foot from the ground to the inside malleolus,	5.0	5.0	5.0	5.2
Height of the foot from the ground to the outside malleolus,	4.0	4.0	4.2	4.5
Length of the foot from the heel to the tip of the big toe,	15.5	15.0	15.8	15.2
The width of the head between the parietal eminences	9.0	9.2	13.0	18.0
The depth of the head from the forehead to the occiput	11.5	11.8	17.0	22.0
Diameter of the neck	6.5	6.5	–	–
Shoulder width from acromion to acromion	23.0	21.5	22.0	25.0
Width of the chest (on a level with the nipples)	17.5	16.0	17.0	21.0
Depth of the chest (on a level with the nipples)	10.5	10.0	12.0	16.0
Width of the hips (over the greater trochanters)	19.5	21.0	18.0	23.0
Width of the thigh at the middle	8.3	8.7	–	–
Width of the leg at the middle of the calf	7.0	6.8	–	–
Width of the foot across the metatarsal-phalangeal articulations	5.7	5.3	5.5	6.5
Width of the arm across the middle of biceps	5.4	5.0	–	–
Width of the forearm across its thickest part	5.9	5.5	–	–
Width of the hand across the metacarpal-phalangeal articulations	5.0	4.5	5.0	6.0

PREPARATORY SWIMMING EXERCISES.

Swimming is an accomplishment that every one should possess, not only because some day one may save his own life or that of another, if he be a skilled swimmer, but also because the art is worth learning for the sake of the exercise itself, which is undoubtedly one of the best for the promotion of physical development. In the summer, the heat usually prevents gymnastic exercises from being practised to any great extent, while swimming will furnish both exercise and recreation, without adding the discomfort of excessive heat. To know how to keep one's self afloat, one need only be familiar with the movements which constitute swimming, and have consciousness of their efficacy.

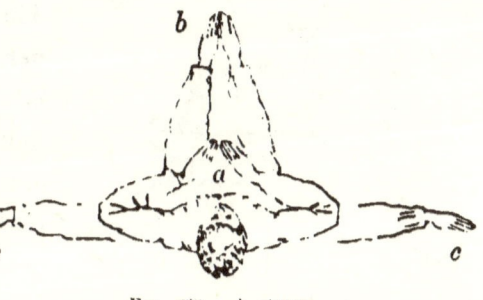

FIG. 239. — A. SWIM.

Consequently, the best plan for acquiring the art is to learn the swimming movements on *terra firma* so thoroughly that we do them unconsciously when we go into the water; and children may be most easily given this practice, if the swimming movements are brought in as part of educational gymnastics.

The exercises are described in accordance with the plan followed in " *Gymnastiska Dagöfningar*," by C. H. Liedbeck, from which hand-book they are borrowed.

St. Introd. to A. Swim. w. even Counts (Fig. 239). — Command, " *Arms forward — bend! Arm swim — one! Two! Three!*". . . 1. The hands are brought so far forward that the fingers touch, the palms facing slightly outward. 2. The arms are stretched forward, the hands still touching and facing as before. 3. The straight arms are moved sideways into yd. pos., while the palms of the hands are turned backward. The arms are horizontal throughout the movement.

Stoop St. Introd. to A. Swim. w. even Counts. — The exercise is done as the previous one, the arms being horizontal throughout the movement.

Stoop St. Introd. to A. Swim. w. uneven Counts. — This is done in a like manner, except that the rhythm is changed, the first two motions being done quickly, and immediately following each other, the third motion being slow. The command is, . . . "*Arm swim — one‿two!! Three!*" . . . These exercises may be brought in among shoulder-blade movements.

FIG. 240. — HANG. INTROD. TO L. SWIM.
a, "ONE!" b, "TWO!" c, "THREE!"

Lying Introd. to L. Swim. w. even Counts (compare Fig. 240). — The pupil is lying face up on a bench with neck firm. Command, "*Leg swim — one! Two! Three!*" . . . 1. With heels held together, the pupil draws his legs up as far as he can, knees and hips bending, and the feet bent upward. 2. Separating the feet, and extending the insteps, the pupil stretches his legs sideways into stride pos. 3. The straight legs are brought together into commencing position.

Lying Introd. to L. Swim. w. uneven Counts. — This is as just described, except that the rhythm is changed, the command being, . . . "*one two! Three!*" . . .

These movements may be used as abdominal exercises. They may also be done in hanging position (Fig. 240) (under hang. or over gr. hang), and are then easy heaving-movements. They may also be done in forward lying position, the pupil grasping the stall-bars, or his hands being held by another pupil.

Str. Forw. Lying Pos. (Fig. 241). — The pupil is lying face down on the floor (on a piece of carpet used for the occasion, or lengthwise on a bench). Command, "*In position — one! Two!*" . . . 1. The pupil swings his arms up into str. pos., and arches his body so that the feet leave the floor.

FIG. 241. — STR. FORW. LYING POS.

2. He resumes commencing position. At first the position may be taken with hips firm instead of arms upward stretch, so as to teach the pupil how to arch the body. The movement is a shoulder-blade movement.

Forw. Ly. A. and L. Swim. — The pupil is lying across a bench (covered by a cushion, if one is at hand). Command. "*Arm and leg swim — one two! Three!*" . . . The movements of the arms and legs are combined and executed on the same principles as above. The exercise is very exhausting, and can be practised only for a few moments at a time; for instance, three or four strokes are done, and then follows a short rest before the movement is repeated.

The progression of these exercises is as follows:

1. St. Introd. to A. Swim. w. even counts.
2. Ly. Introd. to L. Swim. w. even counts.
3. Str. Forw. Ly. Pos.
4. Stoop St. Introd. to A. Swim w. even counts.
5. Stoop St. Introd. to A. Swim. w. uneven counts.
6. Ly. Introd. to L. Swim w. uneven counts.
7. Forw. Ly. A. and L. Swim.

These exercises are best practised in connection with gymnastics, just before the bathing season commences. The movements may then be arranged to occupy two weeks of daily practice, and may be applied according to the following chart (by Liedbeck):

Number of day	1, 2, 3.	4.	5.	6.	7, 8.	9.	10, 11, 12.
Number of exercise	1, 2, 3.	1,2,3,4.	2, 3, 4.	2,3,4,5.	3, 5, 6	5, 6, 7.	7.

The number of exercise refers to the progressive list just given.

After the swimming movements have been practised thus, swimming may be safely tried in shallow water, with the aid of a girdle held by the teacher, or by some other method, the descriptions of which do not belong here.

PHYSIOLOGICAL EFFECTS OF EXERCISE.

On the Lungs. — Every muscular exertion is involuntarily preceded by an increased inhalation, Nature thus providing an increased supply of oxygen to substitute that which is used up by the muscular contraction; in other words, exercise induces respiration; and, proportionately to the force used in the exertion, it also accelerates the respiratory act. In consequence of this, the pulmonary circulation becomes accelerated, and the quantity of air inhaled and of carbon dioxide exhaled is greatly increased. The elimination of water through the lungs also increases in proportion to the exertion.

Dr. Edward Smith's experiments ("Parkes' Practical Hygiene") are interesting. He found that if the quantity of air inhaled in the lying position was taken as unit, that inhaled in sitting position was 1.18; in standing position, 1.33; walking 1 mile an hour, 1.90; walking 4 miles an hour, 5.00; and walking 6 miles an hour, 7.00, etc. Or, in other words, if a man at rest inhales 480 cubic inches of air per minute, while walking 4 miles an hour he inhales 2400 cubic inches, and while running 6 miles an hour 3360 cubic inches. The carbon dioxide exhaled is increased in proportion.

Both the active muscle and the one at rest absorb oxygen and give off carbon dioxide, the absorption of oxygen and exhalation of carbon dioxide in the contracting muscle being about twice as great as in the resting one (Beclard, Helmholz, etc.). If the carbon dioxide were not rapidly carried off by the blood and eliminated from the body, the muscles would soon become unable to work. For it has been proved that, if the pulmonary circulation and the exhalation of carbon dioxide are impeded, muscular exertion soon becomes impossible. Thus, to insure proper elimination of carbon from the body, it is necessary that muscular exercise take place; and when it cannot, the supply of carbon (carboniferous food) must be lessened, if the body is to remain healthy.

Although respiration under ordinary circumstances is an entirely mechanical action, yet it may be controlled by the will, and the latter also governs the respiration unconsciously, since every act of volition will cause it to be hastened. This is true whether the effort be physical or psychical. Respiration is also influenced by temperature, the respiratory act becoming deeper in the same degree as the body loses heat: and, reversely, the temperature of the body will rise as respiration becomes deeper (prolonged respiratory exercises, etc.).

Since oxygen is one of the chief power-producing elements in the body, it can be said that to be strong — capable of much work — good breathing capacity is of greater value than muscular strength; for the former gives one the quality of endurance or ability to persevere, which is the key-note to success in the struggle for the survival of the fittest.

Whereas properly guided exercise will strengthen and develop the breathing apparatus, lack of exercise will weaken the lungs and may even lead to tuberculosis and allied diseases. On the other hand, excessive exercise may cause pulmonary congestion and even hemorrhage from the lungs.

From these facts we gather the following rules to be observed during exercise:

(1) The respiratory organs must be allowed perfect freedom of motion: no tight clothing (corsets!) should be used; and any exercise which compresses the chest or interferes with free respiration is to be avoided.

(2) Since the elimination of carbon dioxide and the need of oxygen are so much increased, the air should be as pure as possible. Hence, exercise out-of-doors is the best; but if it has to be taken indoors, the room should be well ventilated and as far as possible kept free from dust.

(3) The exercises should be so arranged as not to cause overexertion, since the latter produces pulmonary congestion (a condition manifesting itself by deep sighing). Hence, when the exercise causes laborious breathing, rest should take place, or, still better, such movements should be immediately used as will remove this condition (respiratory exercises; slow leg-movements).

(4) Since the elimination of carbon is so much increased, those who exercise should partake of more carboniferous food — best given in the form of fat.

On the Blood and Circulatory Organs. — The increase of color and quickening of the pulse show us that exercise accelerates the circulation.

When a muscle contracts, it exerts a pressure on those vessels which are located in or around it. The arteries have firm and hard walls, and the blood within them flows forward with considerable pressure; furthermore, the semilunar valves at the opening of the aorta prevent the arterial current from taking a backward direction: hence the arteries are but little affected by the pressure of the active muscle. The veins have less elastic walls, and the pressure of the blood in them is less. Their valves preventing the current from flowing in the opposite direction, it follows that the muscular pressure drives the venous current toward the heart with increased speed. At the same time, the *vis a tergo* in the arterial current increases. When the contraction ceases, the veins in question contain less blood. This causes a suction in these vessels, which force, added to the pressure from the arterial side, causes a new supply of blood to rush in with accelerated speed, this being repeated at every contraction and relaxation. At the same time the absorption of oxygen in the contracting part causes the arterial current to flow more quickly to it in order to supply the waste: the *vis a fronte* of the arterial current increases. In fact, active contraction has even more effect on the local afflux than on the return current from the same part.

During flexion and extension, the vessels become alternately shortened and lengthened. Moderately extended, the vessels hold more blood than they do when shortened; hence bodily movements act upon the circulation much after the manner of a "force-pump:" the flow toward the heart is accelerated, and the *vis a tergo* of the blood increased.

Respiration also affects the circulation, as will be seen from the following: At inspiration the elastic air-cells resist the pressure from the inhaled air, and this causes the pressure on other organs (heart and large vessels) within the chest to be less than one atmosphere ("negative pressure"); a suction is caused in direction of the lungs. The blood in the veins is drawn forward to fill the vacuum. The arteries become but little affected, on account of their stiff walls and of the pressure of the blood in them. Besides, the closing of the semilunar valves during the diastole of the ventricles prevents any backward direction of the arterial current. This suction, which by deep inhalations becomes four or five times greater than what it is during ordinary inhalations (Wundt), is a powerful aid in the propulsion of the venous currents. Expiration has the opposite effect: it accelerates the arterial currents, but retards the venous ones.

Every active movement increases the action of the heart, which usually has from ten to thirty more beats per minute during exercise, and sometimes a great deal more. After exercise the heart's action usually falls below the normal — after severe exercise, even as low as fifty to forty per minute. But if due attention is paid to the freedom and extension of the respiratory act, the heart's action need not deviate much from the normal. The same is true, if certain movements are used that will furnish mechanical aid for the propulsion of the blood, *i. e.*, lessen the work for the heart.

Excessive as well as deficient exercise will weaken the heart, and cause diseases, such as palpitation, hypertrophy, dilatation, fatty degeneration, etc. On the other hand, properly guided exercises will tend to strengthen a weakened heart; and in such cases it would be a mistake not to make use of exercise as a therapeutic agent, as well as to suppose that absolute rest is beneficial.

Exercise has also a great influence on the quality of the blood; for, by the increased pressure, the phenomena of endosmosis and exosmosis become livelier, and the absorption of vital constituents becomes more rapid. At the same time, the blood more quickly carries away the waste matter, and throws it off through the organs of excretion; and in this manner its own quality improves.

On the Digestive Organs. — At every inspiration the diaphragm contracts, and exerts a pressure on the contents of the abdomen, the walls of which rise. At expiration the diaphragm rises, and the walls of the abdomen again sink. Thus the diaphragm and the air above it act like a pump, causing a constant oscillation of the abdominal and pelvic contents, increasing the action of their involuntary muscles. This motion aids the peristaltic movements of the stomach and intestine in carrying the food downward, and in hastening the process of digestion.

It is well known how exercise increases the appetite, especially for meat and fat. This increase of appetite indicates a more rapid absorption and a more perfect digestion. On the other hand, lack of exercise lessens the appetite as well as the power of digestion. During exercise, the circulation through the liver is very much accelerated, especially by flexion of the trunk in one direction or another, which movement, by accelerating the circulation through the inferior vena cava, causes a suction in the veins which feed it. A similar effect is caused by deep inspirations.

Finally, if the voluntary muscles of the abdomen are well developed, the fæces are more readily expelled and a livelier peristalsis is insured. Thus exercise, whether general or special, will improve and maintain the normal functional activity of the digestive organs.

On Absorption. — By the increased pressure of the blood, the phenomena of endosmosis and exosmosis become very much livelier, and the flow of the lymph accelerated.

The lymph-vessels are provided with valves like those of the veins — opening toward the heart; hence the alternate contraction and relaxation of adjacent muscles will have the same effect upon the lymphatics as upon the veins: that of forcing the contents of the vessels forward in direction of the heart.

Inhalation has the same effect upon the thoracic duct as upon other vessels in the thoracic cavity: a suction is caused in it which accelerates the flow of its contents. At exhalation, the thoracic duct becomes compressed, and its contents are forced into the venous circulation.

On the Skin. — The cutaneous circulation is accelerated, the skin grows red, and perspiration increases. The amount of evaporating fluid is double (or more) what it is during rest; water, chloride of sodium, acids, and some nitrogen are eliminated. This evaporation lessens the temperature of the body, and prevents it from becoming excessively high. If the pores are closed by dirt, or if there be anything else to check the evaporation, the bodily heat soon rises; the work of the lungs in the elimination of water is greatly increased; breathing becomes laborious, and the power of continuing the exercise rapidly decreases.

After *exertion*, the temperature of the body falls quickly

below the normal degree, while the skin evaporation still continues; hence, danger of taking cold arises.

From this it follows:

(1) That the skin should be kept clean.

(2) That the exercise should be taken in a cool room (55° to 60° F.).

(3) That, during exercise, the clothing worn should be light.

(4) That, after exercise, thicker clothing should be put on to prevent chill.

On the Kidneys. — On account of the increased skin evaporation, not only the water of the urine, but also the chloride of sodium, is lessened. The amount of urea usually remains about the same, although oftentimes it decreases — probably on account of the greater excretion of nitrogen through the lungs (Draper, Regnault, Reiset). Dr. John C. Draper's experiments demonstrate that "after violent exercise, the total amounts of solid residue and urea are diminished, while the proportion of urea to solid residue remains about the same."

On the other hand, by lessening the activity of the skin and lungs, lack of exercise forces the kidneys to increase their excretory function, and may be a contributing cause to various disorders of these organs.

On the Organs of Locomotion. — In active muscles, the temperature increases in proportion to the work (up to a certain limit); there is a change from neutral to alkaline reaction, the principal metamorphoses being the oxidation of the hydrocarbons and the formation of free acid. The venous current from the muscle carries darker blood, less rich in oxygen and water and richer in carbonic acid.

As a result of these changes, the muscles, through regularly taken exercise, grow larger, heavier, and richer in nitrogen;

and as the muscles grow, so do the bones and other passive organs of locomotion. On the other hand, unused muscles diminish in size and strength, and may finally entirely cease to respond to cerebral influence. At the same time, too much exercise of a muscle will cause it to decrease in size and power after it has reached a certain degree of development. This, however, does not occur when all the muscles are moderately exercised, and sufficient time is allowed for rest. If the food does not contain a sufficient amount of nitrogen, other parts of the body have to furnish the supply to the active muscles; and in such a case it is usually the heart that suffers most.

The exhaustion of the oxygen supply and the accumulation of acids in the active muscles soon produce fatigue. In order that the waste may be carried off and a new supply of oxygen stored up, the tired muscle must have a shorter or longer period of rest before contraction can again take place. This alternation between activity and rest occurs even during exercise, although the period of rest at such times is usually insufficient to allow the muscle fully to recuperate. Yet, by letting gymnastic movements follow each other according to physiological principles, the immediate effect of a short period of exercise can be made to be a sense of rest and invigoration instead of fatigue and exhaustion. This, however, cannot be done, when the exercise is prolonged or when it is of a one-sided nature. The general oxygen hunger manifests itself in languor and desire to sleep. Before work can be resumed, all the muscles (and nerves) under the control of the will must be allowed absolute rest until a sufficient amount of oxygen has again been stored up. Artificial means (massage, stimulants, etc.) may occasionally serve instead of rest, but cannot be relied upon as permanent substitutes.

From these facts it follows:

(1) That there is a limit to physical development; and hence that the chief aim of exercise should not be great muscular development, but health.

(2) That this limit is reached sooner, if the exercise is of a one-sided character.

(3) That during periods of exercise, the food should contain more nitrogen.

(4) That, if good physical development is to be acquired, the exercise should include the whole body, and should alternate with periods of sufficient rest.

On the Nervous System. — Experience has proved that inactive motor nerves grow weaker, become fattily degenerated, and finally lose all irritability, so that it becomes impossible even to think in the direction of movements to which those nerves should give the impulse. At the same time the sensory nerves become highly sensitive and even morbidly irritable. The nervously unstrung are to be sought among brain-workers of sedentary habits, — not among laborers or persons leading an active out-of-door life.

In all active exercise, the cerebral influence to the acting parts is highly increased; and it might be said that the nerves in question are even more active than the muscles which they move. The consequence of this is, that the nutritive results of muscular contractions are taken up by the nerves still more than by the muscles. Thus exercise becomes efficient in strengthening and developing the nervous system, and in maintaining it in normal condition. In fact, exercise may often serve as a valuable remedial agent in nervous disorders.

On the Mind. — As the general circulation and the quality of the blood improve, the brain becomes better nourished, and its power of action increases.

It is only in a healthy body that we find a healthy mind; whereas, if the body is weakened by disease, — by inactivity, — the intellectual powers become enfeebled.

Many maintain that it is not possible to combine great mental work with powerful bodily exercise, but practical experiments have proved this to be an entirely mistaken idea; for, if the two are made judiciously to alternate, it will be found that the results of each become much more extensive, much better in quality. If athletes are stupid, it is because they make no efforts toward mental cultivation, and not because exercise lessens their powers in this direction.

The moral forces are also affected by exercise, for immorality goes hand in hand with morbidity of mind. If the latter can be changed, the former will also improve. It is the weak, and not those who are strong in body and mind, that recruit the jails.

Exercise develops a consciousness of power, which inspires courage, confidence, and resolution. Through its influence the moral self comes forth healthier, purer, and stronger, and man becomes in every way better fitted to lead a life of usefulness to his fellow-men.

MEMORANDA OF THE MUSCULAR ACTIVITY IN THE PRINCIPAL MOVEMENTS OF THE HUMAN BODY

MEMORANDA OF THE MUSCULAR ACTIVITY IN THE PRINCIPAL MOVEMENTS OF THE HUMAN BODY.

THE HEAD AND NECK.

MOVEMENT.	MUSCLE.	ORIGIN.	INSERTION.
Forward flexion.	Platysma Myoides.	Clavicle. Acromion. Fascia of { Pectoral, Deltoid, Trapezius. }	Lower jaw. Angle of the mouth. Cellular tissue of the face.
	Sterno-cleido-mastoid.	Sternum. Inner third of clavicle.	Mastoid process of temporal bone. Superior curved line of occipital bone.
	Rectus Capitis Anticus Major.	Anterior tubercles, transverse processes of 3d to 6th cervical vertebræ.	Basilar process of occipital bone.
	Rectus Capitis Anticus Minor.	Anterior lateral surface of atlas. Transverse process of atlas.	Basilar process of occipital bone.
	Longus Colli.	Anterior tubercles, transverse processes of 3d, 4th, and 5th cervical vertebræ. Bodies of first 3 dorsal vertebræ. Bodies of lower 3 cervical and upper 3 dorsal vertebræ.	Tubercle, anterior arch of atlas. Transverse processes, 5th and 6th cervical vertebræ. Bodies of 2d, 3d, and 4th cervical vertebræ.
	Scalenus (a. Anticus, b. Medius, c. Posticus).	a, b. Upper surface of 1st rib. c. Outer surface of 2d rib.	a. Transv. proc., 3d to 6th cervic. verteb. b. Transv. proc., lower 6 cervic. verteb. c. Posterior tubercle, transverse processes, lower three cervical vertebræ.
	Digastric.	Mastoid process of temporal bone. Inner surface of inferior maxillary.	Hyoid bone. Inferior, inner side of chin.
	Sterno-hyoid.	Posterior surface of sternum. Sternal end of clavicle.	Body of hyoid bone.
	Sterno-thyroid.	Posterior surface of sternum. Cartilage of 1st rib.	Oblique line of thyroid cartilage.

APPENDIX

	Muscle	Origin	Insertion
	Omo-hyoid.	Upper border of scapula. Transverse scapular ligament.	Body of hyoid bone.
	Mylo-hyoid.	Mylo-hyoid ridge of lower jaw.	Body of hyoid bone.
	Genio-hyoid.	Genial tubercle of inferior maxillary.	Body of hyoid bone. Side of pharynx. Whole inferior surface of the tongue.
Backward flexion.	Trapezius.	Occipital bone, inner third of superior curved line. Ligamentum nuchæ. Spinous processes of last cervical and all dorsal vertebræ. Supra-spinous ligament.	Posterior border of clavicle, outer third. Acromion, superior margin. Spine of scapula, superior border.
	Levator Anguli Scapuli	4 upper cervical vertebræ, transverse proc.	Scapula, posterior border.
	Splenius.	Ligamentum nuchæ, lower half. Last cervical and 6 upper dorsal spinous processes.	Mastoid process. Occipital bone, outer third of superior curved line. 3 or 4 upper cervic. verteb., transv. proc.
	Cervicalis Ascendens.	Angles of 4 or 5 upper ribs.	4th, 5th, and 6th cervic. verteb., transv. proc.
	Transversalis Colli.	3d, 4th, 5th, and 6th dorsal vertebræ, transverse processes.	Lower 5 cervical verteb., transv. proc.
	Trachelo-mastoid.	3d, 4th, 5th, and 6th dorsal vertebræ, transverse processes. 3 or 4 lower cervic. verteb., articular proc.	Mastoid process, posterior margin.
	Spinalis Colli.	5th and 6th cervic. verteb., spinous processes. (1st and 2d dorsal verteb., spinous processes.)	Spine of axis. (Spine of 3d and 4th cervical vertebræ.)
	Complexus.	Last cervical and 3 upper dorsal vertebræ, transverse processes. 4th, 5th, and 6th cervic. verteb., articular proc.	Occipital bone, between the curved lines.
	Multifidus Spinæ.	Back of sacrum. Ilium, posterior superior spine. Posterior sacro-iliac ligaments. Cervical vertebræ, articular processes. Lumbar and dorsal verteb., transv. proc.	Inferior border and lateral surface of the transverse processes of all true vertebræ, except the first one.
	Inter-spinales.	Between spinous processes, 6 pairs in cervical, 3 in dorsal, and 4 to 6 in lumbar region.	

MOVEMENT.	MUSCLE.	ORIGIN.	INSERTION.
Backward flexion.	Rectus Capitis Posticus Minor.	Posterior arch of atlas.	Occipital bone, inner third of inferior curved line.
	Obliquus Capitis Superior.	Transverse process of atlas.	Occipital bone, between the curved lines.
Sideways flexion.	Platysma Myoides, Sterno-cleido-mastoid, Trapezius, Splenius, Trachleo-mastoid, Complexus, Scalenus.	See above.	See above.
	Rectus Capitis Posticus Major.	Spine of axis.	Occiput, middle third of inferior curved line.
	Inter-transversales.	Between transverse processes of all true vertebræ, except last lumbar.	
	Rectus Lateralis.	Transverse process of atlas, upper surface.	Occipital bone, jugular process.
Rotation.	Complexus, Multifidus Spinæ, Cervicalis Ascendens, Splenius, Trachleo-mastoid, Sterno-cleido-mastoid.	See above.	See above.
	Obliquus Capitis Inferior.	Spinous process of axis.	Transverse process of atlas.

THE TRUNK.

MOVEMENT.	MUSCLE.	ORIGIN.	INSERTION.
Forward flexion.	Rectus Abdominis.	Pubic crest. Symphysis pubis.	Cartilage of 5th, 6th, and 7th ribs.
	Pyramidalis.	Front of pubic bone. Anterior pubic ligament.	Linea alba, half-way to umbilicus.
	External Oblique.	Digitations from 8 lower ribs.	Crest of ilium, anterior outer lip. Linea alba. Ensiform cartilage. Symphysis and crista pubis. Pectineal line.

	Muscle	Origin	Insertion
	Internal Oblique.	Lumbar fascia. Crest of ilium, anterior two-thirds of middle lip. Poupart's ligament, outer half.	Cartilage of 3 lower ribs. Linea alba, by aponeurosis. Pubic crest, } by tendon of transversalis. Pectineal line, }
	Psoas Magnus.	Bodies and transverse process of last dorsal and all lumbar vertebræ.	Femur, lesser trochanter.
	Psoas Parvus.	Bodies of last dorsal and first lumbar verteb.	Ilio-pectineal eminence. Iliac fossa.
Forward downward flexion.	*The above muscles, assisted by* Pectoralis Major,		
	Pectoralis Minor,	Clavicle, sternal one-half. Front of sternum, down to the 6th or 7th rib. Cartilages of all true ribs. External oblique.	Humerus, bicipital ridge.
		3d, 4th, and 5th ribs. Intercostal aponeurosis.	Coracoid process of scapula.
	Serratus Magnus,	9 digitations from 8 upper ribs (2 on the 2d). Intercostal aponeurosis.	Inner margin, posterior border of scapula.
	Transversalis,	Poupart's ligament, outer one-third. Crest of ilium, anterior two-thirds of inner lip. Cartilages of 6 lower ribs. Aponeurosis from spinous and transverse processes of lumbar vertebræ.	Linea alba. With internal oblique into pubic crest and pectineal line.
	Pectineus,	Ilio-pectineal line and crista pubis. Gimbernat's ligament.	Femur, between trochanter minor and linea aspera.
	Adductor Longus,	Front of pubes.	Linea aspera, middle third.
	Adductor Brevis,	Descending ramus of pubes.	Linea aspera, upper third.
	Rectus Femoris,	Ilium, anterior, inferior spine. Groove above the brim of acetabulum.	Over patella into tuberosity of tibia.
	Sartorius.	Ilium, anterior, superior spine. Iliac notch, superior one-half.	Upper inner surface, shaft of tibia, near the crest.
Backward flexion.	Trapezius.	See above.	See above.
	Rhomboideus.	Ligamentum nuchæ. Spines of last cervic. and first 4 dorsal verteb. Supra-spinous ligament.	Inner border of scapula.

MOVEMENT.	MUSCLE.	ORIGIN.	INSERTION.
Backward flexion.	Latissimus Dorsi.	Aponeurosis from sacral, lumbar, and 5 lower dorsal vertebræ. Crest of ilium, posterior one-fourth. Digitations from the 3 or 4 lower ribs.	Humerus, bicipital groove.
	Serratus Posticus Superior.	Ligamentum nuchæ. Spines of last cervical and 2 upper dorsal vert.	4 digitations into upper outer border of 2d to 5th ribs inclusive.
	Serratus Posticus Inferior.	Spines of last 2 dorsal and first 3 lumbar verteb.	4 digitations into lower border of the 4 lower ribs.
	Erector Spinæ.	Sacro-iliac groove. Sacral, lumbar, and 3 lower dorsal spines. Iliac crest. Posterior eminences of sacrum.	The next two muscles.
	Sacro-lumbalis.	Erector spinæ.	Angles of 6 lower ribs.
	Longissimus Dorsi.	Erector spinæ.	Transverse processes of lumbar vertebræ. Transverse processes of all dorsal vertebræ. Between tubercles and angles of 7th to 11th ribs.
	Spinalis Dorsi.	Spines of last 2 dorsal and first 2 lumbar vertebræ.	Spines of 2d to 8th dorsal vertebræ.
	Semispinalis Dorsi.	Digitations from transverse processes of 6 lower dorsal vertebræ.	Spines of upper 4 dorsal and lower 2 cervical vertebræ.
	Multifidus Spinæ.	See above.	See above.
	Quadratus Lumborum.	Ilio-lumbar ligament. Posterior portion of inner crest of ilium. Transverse proc. of lower 3 or 4 lumbar verteb.	Transverse processes of 3d, 4th, and 5th lumbar vertebræ. Lower border of 12th rib, posterior one-half.
	Levatores Costarum.	Transverse process of a dorsal vertebra.	Next rib below, upper surface near the angle.
Sideways flexion.	External Oblique, Internal Oblique, Quadratus Lumborum, Longissimus Dorsi, Sacro-lumbaris.	See above.	See above.

APPENDIX

SHOULDER-BLADE.

MOVEMENT.	MUSCLE.	ORIGIN.	INSERTION.
Rotation.	The oblique and transverse muscles of each opposite lateral half of the trunk (and upper leg).		
Drawn backward and downward.	Latissimus Dorsi, Rhomboideus; Trapezius (middle portion).	See above.	
Drawn forward and downward.	Serratus Magnus, Pectoralis Minor.	See above.	
Elevation.	Levator Anguli Scapula, Trapezius (superior portion), Rhomboideus.	See above.	
Depression.	Trapezius (inferior portion), Latissimus Dorsi, Pectoralis Minor.	See above.	

HUMERUS.

MOVEMENT.	MUSCLE.	ORIGIN.	INSERTION.
Elevation sideways to horizontal position.	Deltoid.	Clavicle, outer one-third. Acromion, outer margin and upper surface. Spine of scapula, lower border.	Shaft of humerus, outer surface near its middle.
	Supraspinatus.	Supra-spinous fossa, inner two-thirds.	Greater tuberosity of humerus, superior facet.

MOVEMENT.	MUSCLE.	ORIGIN.	INSERTION.
Sideways from horizontal to vertical position.	Serratus Anticus Major. Trapezius.	See above.	
Elevation forward, upward.	Deltoid (anterior portion). Pectoralis Major.	See above.	
	Coraco-brachialis.	Apex of coracoid process.	Inner side of the shaft of humerus, near its middle.
	Biceps (long head).	Upper margin of glenoid cavity.	Tuberosity of radius. Fascia of forearm.
Elevation backward.	Deltoid.	See above.	
	Teres Major.	Scapula, inferior angle.	Bicipital ridge of humerus.
	Teres Minor.	Axillary border of scapula, upper two-thirds.	On and below the greater tuberosity of humerus.
	Triceps (long head).	Depression below the glenoid cavity of scapula.	Upper end of olecranon process.
	Latissimus Dorsi.	See above.	
Depression from horizontal position.	Pectoralis Major, Latissimus Dorsi, Teres Major and Minor.	See above.	
	Subscapularis.	Sub-scapular fossa, inner two-thirds.	Lesser tuberosity of humerus.
Rotation inward.	Anterior portion of Deltoid; Subscapularis, Pectoralis Major, Latissimus Dorsi, Teres Major.	See above.	
Rotation outward.	Posterior portion of Deltoid; Supraspinatus, Teres Minor, Coraco-brachialis.	See above.	
	Infraspinatus.	Infra-spinous fossa, inner two-thirds.	Greater tuberosity of humerus.

FORE-ARM.

MOVEMENT.	MUSCLE.	ORIGIN.	INSERTION.
Flexion.	Biceps.	Upper margin of glenoid cavity. Apex of coracoid process.	Tuberosity of radius. Fascia of forearm.
	Brachialis Anticus.	Lower half of the shaft of humerus, anterior lateral aspect.	Coronoid process of ulna.
	Pronator Teres.	Above inner condyle of humerus. Lower portion of inter-muscular septum. Inner side of coronoid process of ulna.	Shaft of radius, outer side near its middle.
	Supinator Longus.	External condyloid ridge of humerus, upper two-thirds. Intermuscular septum.	Styloid process of radius.
	Flexor Carpi Radialis.	Inner condyle. Intermuscular septum.	Base of 2d metacarpal bone.
	Palmaris Longus.	Same as Flexor Carpi Radialis.	Annular ligament. Palmar fascia.
	Flexor Carpi Ulnaris.	a. Inner condyle. b. { Inner margin of olecranon. Upper two-thirds, posterior border of ulna. Intermuscular septum. }	Pisiform bone. Annular ligament. Base of 5th metacarpal bone.
	Flexor Sublimus Digitorum.	a. { Inner condyle. Internal lateral ligament. Intermuscular septum. } b. Inner side of coronoid process. c. Oblique line of radius.	4 tendons into lateral margins of the second phalanges.
Extension.	Triceps.	a, b. Posterior surface, shaft of humerus above and below the musculo-spinal groove. c. Depression below the glenoid cavity of scapula.	Upper end of olecranon process.
	Subanconeus.	Humerus, above olecranon fossa.	Posterior ligament of elbow-joint.

MOVEMENT.	MUSCLE.	ORIGIN.	INSERTION.
Rotation inward.	Pronator Teres, Flexor Carp. Rad., Palm. Long., Flex. Subl. Digit.	See above.	
	Pronator Quadratus.	Oblique line, lower one-fourth of ulna. Internal border of ulna.	Shaft of humerus, lower one-fourth, anterior surface and external border.
Rotation outward.	Biceps.	See above.	
	Supinator Brevis.	External condyle of humerus. External lateral and orbicular ligaments. Oblique line of ulna.	Anterior superior surface of radius.
	Extensor Secundi Interaodii Pollicis.	Posterior surface, shaft of ulna. Interosseous membrane.	Base of last phalanx of the thumb.

HAND.

MOVEMENT.	MUSCLE.	ORIGIN.	INSERTION.
Flexion.	Flexor Carpi Radialis and Ulnaris, Palmaris Longus, Flex. Subl. Digitorum.	See above.	
	Flexor Profundus Digitorum.	Upper two-thirds, shaft of ulna. Depression on inner side of coronoid process. Interosseous membrane.	4 tendons to bases of last phalanges.
	Flexor Longus Pollicis.	Upper two-thirds, shaft of radius. Interosseous membrane.	Base of last phalanx of thumb.
Extension.	Extensor Carpi Radialis Longior.	Lower one-third, external condyloid ridge of humerus. Intermuscular septum.	Radial side, base of 2d metacarpal bone.
	Extensor Carpi Radialis Brevior.	External condyle. External lateral ligament. Intermuscular septa.	Radial side, base of 3d metacarpal bone.

APPENDIX

MOVEMENT	MUSCLE	ORIGIN	INSERTION
	Extensor Carpi Ulnaris.	Middle one-third, posterior border of ulna. Fascia of forearm.	Base of 5th metacarpal bone.
	Extensor Secundi Internodii Pollicis.	See above.	
	Extensor Indicis.	Shaft of ulna, posteriorly. Interosseous membrane.	2d and 3d phalanges of index finger.
	Extensor Communis Digitorum.	External condyle. Deep fascia. Intermuscular septa.	3 tendons into 2d and 3d phalanges of all the fingers.
	Extensor Minimi Digiti.	Same as the one just named.	2d and 3d phalanges of the little finger.
Flexion outward.	Flex. Carpi Rad., Extensor Carp. Rad. Longior and Brevior.	See above.	
	Extensor Ossis Metacarpi Pollicis.	Posterior surface, shafts of radius and ulna. Interosseous ligament.	Base of 1st metacarpal bone.
	Extensor Primi Internodii Pollicis.	Posterior surface, shaft of radius. Interosseous membrane.	Base of 1st phalanx of the thumb.
Flexion inward.	Ext. Carpi Uln., Flex. Carpi Uln., Subl. and Prof. Digit.	See above.	

THIGH.

MOVEMENT	MUSCLE	ORIGIN	INSERTION
Flexion (elevation forward).	Psoas Magnus.	Iliac fossa. Inner margin, iliac crest. Ilio-lumbar ligament. Base of sacrum.	
	Iliacus.	Anterior spinous processes of ilium, and notch between them. Capsule of hip-joint.	Side of the tendon of psoas magnus.

MOVEMENT.	MUSCLE.	ORIGIN.	INSERTION.
Flexion (elevation forward).	(Sartorius and) Adductors; Pectineus;	See above.	
	Gracilis.	Inner margin, rami of pubes and ischium.	Shaft of tibia, below the inner tuberosity.
	Gluteus Minimus.	Ilium, between the middle and inferior curved lines.	Anterior border of the great trochanter.
	Obturator Externus.	Margin of the great sacro-sciatic notch. Inner margin of obturator foramen. Anterior surface of obturator membrane.	Digital fossa, neck of femur.
	(Tensor Vaginæ Femoris.)	Crest of ilium, anterior part of outer lip. Anterior superior spinous processes.	Fascia lata.)
	(Rectus Femoris.)	See above.)	
Extension.	Gluteus Maximus.	Superior curved line and crest of ilium. Last piece of sacrum. Side of coccyx. Great and posterior sacro-sciatic ligaments.	Fascia lata. Rough line between great trochanter and linea aspera.
	Gluteus Medius.	Ilium, between superior and middle curved lines. Fascia of the same part.	Oblique lines on great trochanter.
	Pyriformis.	3 digitations from front of sacrum. Margin of great sacro-sciatic foramen. Great sacro-sciatic ligament.	Upper border of great trochanter.
	Obturator Internus.	Posterior margin of obturator foramen. Inner surface of obturator membrane.	Great trochanter.
	Gemellus { Superior, Inferior.	Spine of ischium, outer surface. Tuberosity of ischium.	Great trochanter and tendon of obturator internus.
	Quadratus Femoris.	Tuberosity of ischium.	Linea quadrati between the trochanters.
	Biceps Femoris (long head).	Tuberosity of ischium.	Outer side, head of fibula.
	Adductor Magnus (posterior portion).	Rami of ischium and pubes. Tuberosity of ischium, outer margin, under surface.	Rough line between the great trochanter and linea aspera. Linea aspera. Tubercle above inner condyle of femur.

MOVEMENT.	MUSCLE.	ORIGIN.	INSERTION.
Abduction.	Semi-tendinosus.	Tuberosity of Ischium (tendon in common with Biceps Femoris). Adjacent aponeurosis.	Upper inner surface, shaft of tibia.
	Semi-membranosus.	Tuberosity of Ischium.	Inner tuberosity of tibia.
	The 3 Glutei, Tensor Vag. Femoris, Pyriformis, Gemelli, Sartorius, and Obturator Internus.	See above.	
Adduction.	The 3 Adductors, Psoas Magnus, Iliacus, Pectineus, Quadratus Fem., Obt. ext., Gracilis, Semi-tend. and membr., Long head of Biceps.	See above.	
Rotation inward.	Tensor Vag. Fem., Gluteus Medius (anterior fibres) and Minimus, Semi-tend. and membr., Gracilis, Sartorius.	See above.	
Rotation outward.	Pyriformis, Obt. int. and ext., Gemelli, Quadratus Fem., Gluteus Maximus and Medius (posterior fibres), Psoas Mag., Iliacus Pectineus, the 3 Adductors, Biceps.	See above.	

LEG.

MOVEMENT.	MUSCLE.	ORIGIN.	INSERTION.
Flexion.	Biceps, Semi-tend., Semi-membr., Gracilis, Sartorius.	See above.	
	Gastrocnemius.	Condyles of femur. Supra-condyloid ridges. } 2 heads.	By tendo achillis into posterior tuberosity of os calcis.

MOVEMENT.	MUSCLE.	ORIGIN.	INSERTION.
Flexion.	Plantaris.	Outer bifurcation of linea aspera. Posterior ligament of knee-joint.	Os calcis, posterior surface.
	Popliteus.	External condyle of femur. Posterior ligament of knee-joint.	Shaft of tibia, triangular surface above oblique line.
Extension.	Rectus Femoris.	See above.	
	Vastus Externus.	Great trochanter, anterior border. Whole length of linea aspera.	Tuberosity of tibia.
	Vastus Internus and Crureus.	Linea aspera, inner lip. Shaft of femur, nearly whole length, anteriorly and laterally.	Tuberosity of tibia.
Rotation outward. (Flexed leg.)	Biceps and Gastrocnemius.	See above.	
Rotation inward. (Flexed leg.)	Semi-tend. and membr., Sartorius, Gracilis, Popliteus, Plantaris, Gastrocnemius.		

FOOT.

MOVEMENT.	MUSCLE.	ORIGIN.	INSERTION.
Flexion.	Tibialis Anticus.	Shaft of tibia, outer tuberosity, and upper two-thirds. Interosseous membrane. Deep fascia. Intermuscular septum.	Inner plantar surface of internal cuneiform bone. Base of 1st metatarsal bone.
	Extensor Proprius Hallucis.	Middle two-fourths of fibula, anteriorly. Interosseous membrane.	Base of last phalanx of the great toe.

	Muscle	Origin	Insertion
	Extensor Longus Digitorum.	Outer tuberosity of tibia. Shaft of fibula, upper three-fourths, anteriorly. Interosseous membrane. Deep fascia. Intermuscular septa.	2d and 3d phalanges of the 4 lesser toes.
	Peroneus Tertius.	Outer, lower one-fourth of fibula. Interosseous membrane. Intermuscular septum.	Base of 5th metatarsal bone.
Extension.	Gastrocnemius and Plantaris.	See above.	
	Soleus.	Shaft of fibula, head and upper half, posteriorly. Tibia, oblique line. Tendinous arch.	Tendo Achillis.
	Flexor Longus Hallucis.	Shaft of fibula, lower two-thirds, internally. Interosseous membrane. Fascia. Intermuscular septum.	Great toe, base of last phalanx.
	Flexor Longus Digitorum.	Shaft of tibia, posteriorly, below the oblique line.	4 lesser toes, bases of last phalanges.
	Tibialis Posticus.	Shaft of tibia, upper one-half, posteriorly. Shaft of fibula, upper two-thirds, internally. Interosseous membrane. Deep fascia. Intermuscular septum.	Tuberosity of scaphoid bone. Internal cuneiform bone. Base of 1st metatarsal bone.
	Peroneus Longus.	Head of fibula and upper two-thirds of shaft. Deep fascia. Intermuscular septa.	Base of 1st metatarsal bone, outer side.
	Peroneus Brevis.	Shaft of fibula, middle one-third, externally. Intermuscular septa.	Base of 5th metatarsal bone, dorsal side.
Rotation outward.	Peroneus Longus, Brevis and Tertius (Extensor Longus Digitorum).	See above.	
Rotation inward.	Tibialis Anticus and Posticus, Flexor Longus Digitorum, Flexor Longus Hallucis.	See above.	

SET No. 1.

Twenty-five tables of exercises for a class of children, eight to ten years of age, the exercises to occupy about fifteen minutes a day. No apparatus to be used.

I.
1. Fund. pos.
2. Close st. pos.
3. Facings.
4. Wing st. pos.
5. Stride st. pos.
6. Bend st. pos.
7. Yard *a* st. pos.

II.
1. St. heel-elev.
2. Walk st. *b* pos.
3. St. 2 A. ext. sidew.
4. Wing st. T backw. flex.
5. Stride st. heel-elev.
6. Yard *a* st. 2 A. fling.

III.
1. Close st. heel-elev.
2. St. II. backw. flex.
3. Wg. st. T. backw. flex.
4. St. 2 A. ext. upw. (also sidew.).
5. Wg. stride st. sidew. flex.
6. Yd. *a* st. 2 A. fling.

IV.
1. Walk st. *b* heel-elev.
2. St. 2 A. ext. upw.
3. Wg. stride st. T. backw. flex.
4. Slow march on tip-toe.
5. Wg. close st. T. rot.
6. See III. 6.

V.
1. Walk st. *b* pos. in series.
2. Half str. st. pos.
3. Wg. walk st. *b* backw. flex.
4. St. 2 A. ext. forw.
5. See IV. 4.
6. Wg. stride st. T. rot.
7. St. 2 A. elev. sidew.

VI.
1. Walk st. *a* pos.
2. H. flex. sidew.
3. A. fling. to half str. st. pos.
4. Wg. turn close st. T. backw. flex.
5. Wg. st. 2 Kn. flex.
6. See V. 6.

VII.
1. Str. stride st. pos.
2. St. 2 A. fling. upw.
3. Wg. turn. stride st. backw. flex.
4. Half str., half yd. st. pos.
5. Wg. st. 2 Kn. flex.
6. Wg. close st. sidew. flex.
7. Yd. *d* st. 2 A. elev.

VIII.
1. Str. stride st. heel-elev.
2. Rest st. T. backw. flex.
3. Bend st. A. ext. upw.
4. Wg. st. 2 Kn. flex.
5. St. 2 A. ext. backw.
6. Wg. stride st. T. rot.
7. See VII. 7.

IX.
1. Str. walk st. *a* heel-elev.
2. St. 2 A. ext. forw. sidew. and upw.
3. Rest stride st. T. backw. flex.
4. Half str., half reach st. pos.
5. Yd. st. 2 Kn. flex.
6. Rest stride st. sidew. flex.
7. St. 2 A. elev. forw. upw.

X.
1. Str. walk st. *b* heel-elev.
2. St. 2 A. ext. forw. sidew. and upw.
3. Rest stride st. T. backw. flex.
4. Yd. stoop stride st. H. rot.
5. Half str. close st. sidew. flex.
6. Wg. st. prep. to jump.
7. See IX. 7.

XI.

1. Str. walk st. c heel-elev.
2. Wg. turn. stride st. backw. flex.
3. Half yd., half reach st. pos.
4. Wg. courtesy sitt. pos.
5. Wg. walk st. *a* T. rot.
6. Wg. st. alt. Kn. flex. upw.
7. St. 2 A. elev. sidew. upw.

XII.

1. Wg. st. prep. to jump.
2. Wg. st. H. rot. w. flex.
3. Wg. turn walk st. *a* backw. flex.
4. Yd. stoop stride st. 2 A. fling.
5. Yd. courtesy sitt. pos.
6. Half str. walk st. c sidew. flex.
7. See XI. 7.

XIII.

1. Wg. walk st. *b* 2 Kn. flex.
2. Rest turn. close st. backw. flex.
3. Yd. *d* st. alt. A. flex.
4. Wg. cr. half st. pos.
5. Wg. walk st. *b* T. rot.
6. Yd. st. prep. to jump.
7. Yd. walk st. *b* 2 A. fling. and change of feet.

XIV.

1. Str. walk st. *b* 2 Kn. flex.
2. St. 2 A. ext. forw., etc.
3. Rest turn. close st. backw. flex.
4. Wg. cr. half st. F. flex. and ext.
5. Bend stoop stride st. alt. A. ext. upw.
6. Rest close st. sidew. flex.
7. See XIII. 7.

XV.

1. Wg. toe st. H. rot.
2. St. 2 A. ext. in various dir.
3. Str. stride st. backw. flex.
4. Str. stride st. 2 Kn. flex.
5. Yd. stoop stride st. 2 A. fling.
6. Wg. walk st. c T. rot.
7. St. 2 A. elev. sidew. w. heel-elev.

XVI.

1. Rest st. 2 Kn. flex.
2. Str. stride st. T. flex. backw., forw. and downw.
3. Wg. cr. half st. Kn. ext. backw.
4. Bend st. alt. A. ext. forw.
5. Rest walk st. c sidew. flex.
6. Upw. jump.
7. See XV. 7.

XVII.

1. Yd. toe st. 2 Kn. flex. to sitt. pos.
2. Rest turn. walk st. *a*. backw. flex.
3. 2 A. ext. in various dir.
4. Str. stride st. T. rot.
5. Yd. st. 2 A. fling. forw.
6. Upw. jump.
7. Yd. *d* st. 2 A. and heel-elev.

XVIII.

1. Half str., half yd., etc., pos.
2. See XVII. 2.
3. Str. stride st. 2 Kn. flex.
4. Str. close st. T. rot.
5. See XVII. 5.
6. 90° upw. jump.
7. See XVII. 7.

XIX.

1. Wg. courtesy sitt. H. rot.
2. Str. st. backw. forw. down. flex.
3. Reach st. 2 A. fling. upw.
4. Wg. cr. half st. Kn. ext. forw.
5. Bend turn close st. 2 A. ext. upw.
6. 90° upw. jump.
7. 2 A. elev. forw. upw. w. heel-elev.

XX.

1. Str. walk st. *a* 2 A. ext. and ch. F.
2. Rest turn stride st. backw. flex.
3. Str. stride courtesy sitt. pos.
4. Wg. turn close st. sidew. flex.
5. Str. stoop stride st. 2 A. ext. sidew. and upw.
6. 180° upw. jump.
7. See XIX. 7.

XXI.
1. Yd. stride st. 2 A. ext. and ch. of F.
2. Rest turn stride st. backw. flex.
3. ½ str. ¼ yd. stoop stride st. pos.
4. Rest cr. ½ st. Kn. ext. backw.
5. Bend turn stride st. 2 A. ext. upw.
6. 180° upw. jump.
7. 2 A. elev. sidew. upw. and heel-elev.

XXII.
1. Wg. stride courtesy sitt. pos.
2. Str. walk st. *b* T. flex. backw. forw. and downw.
3. Wg. st. L. elev.
4. Reach stoop stride st. 2 A. fling.
5. Wg. turn walk st. *a* sidew. flex.
6. With 2 A. fling upw. jump.
7. See XXI. 7.

XXIII.
1. Wg. st. H. rot. w. flex.
2. Bend arch stride st. 2 A. ext. upw.
3. Wg. courtesy st. H. rot.
4. See XXII. 4.
5. Str. stride st. sidew. flex.
6. See XXII. 6.
7. Yd. *d* toe st. 2 A. clev. w. 2 Kn. flex.

XXIV.
1. Rest stride courtesy sitt. pos.
2. See XXIII. 2.
3. Bend toe st. 2 A. ext. w. 2 Kn. flex.
4. Yd. stoop stride st. 2 A. clev.
5. Rest turn close st. sidew. flex.
6. W. 2 A. and L. fling upw. jump.
7. Yd. turn st. 2 A. fling.

XXV.
1. Bend courtesy sitt. 2 A. ext. upw. and sidew.
2. See XXIII. 2.
3. Rest st. L. elev.
4. Yd. stoop stride st. 2 A. clev.
5. Str. stride st. quick T. rot.
6. See XXIV. 6.
7. Yd. turn stride st. 2 A. fling.

The balance-movement is occasionally exchanged for slow march on tip-toe with the arms in rest or str. pos. If the time allows, other movements — for rhythm and order, etc. — are added, whenever the teacher considers them needed.

SET No. 2.[1]

Twenty-five tables of exercises for a class of children twelve to fourteen years of age, who have had previous training in gymnastics. The school-desks and chairs to be used as apparatus. The exercises to occupy about twenty minutes a day.

I.
1. { St. H. backw. flex.
   { Wg. st. heel-elev.
2. St. 2 A. ext. sidew.
3. Wg. st. 2 Kn. flex.
4. Wg. st. T. backw. flex.
5. ¼ str. close st. T. sidew. flex.
6. Prep. to jump.
7. St. 2 A. elev.

II.
1. { St. H. sidew. flex.
   { Wg. stride st. heel-elev.
2. St. 2 A. ext. upw.
3. Wg. st. alt. Kn. flex. upw.
4. Wg. turn close st. T. backw. flex.
5. ¼ str. yd. st. pos.
6. Rest stride st. T. sidew. flex.
7. Prep. to jump.
8. Yd. *a* st. 2 A. fling.

[1] This set, as well as all the others, represent a great deal more than a year's work; for, in most cases, the progression will have to be slower than what the tables indicate.

III.

1. { St. 2 A. elev.
   { Wg. st. 2 Kn. flex.
2. St. 2 A. ext. forw. and sidew.
3. Wg. (later str.) stride st. 2 Kn. flex.
4. Wg. turn wlk. st. *a* T. backw. flex.
5. Yd. stoop stride st. H. rot.
6. Rest close st. T. sidew. flex.
7. Upw. jump.
8. See II. 8.

IV.

1. { Wg. wlk. st. *a* heel-elev.
   { ½ str. reach st. pos.
2. Rest st. T. backw. flex. (forw. flex. in str. pos.)
3. St. 2 A. ext. forw. sidew. upw. and downw.
4. Wg. (yd.) courtesy sitt. H. rot.
5. 2 A. ext. sidew. and backw.
6. ½ str. wlk. st. *c* T. sidew. flex.
7. Upw. jump.
8. Yd. *d* st. 2 A. elev.

V.

1. { Wg. wlk. st. *c* heel-elev. (later in 4 counts).
   { Bend st. alt. A. ext. upw.
2. Rest stride st. T. backw. flex. (forw. flex. in str. pos.)
3. Fall hang. pos. (Facing the back of the room, the pupil grasps the upper front edges of the two desks and swings his feet forward in the aisle.)
4. Str. wlk. st. *b* 2 Kn. flex.
5. Yd. stoop stride st. 2 A. fling.
6. Str. close st. T. rot. (later bend turn close st. 2 A. ext.).
7. 90° upw. jump.
8. St. 2 A. elev. forw. upw.

VI.

1. { Wg. wlk. st. *b* heel elev. (later in 4 counts).
   { Yd. *d* st. alt. A. flex.
2. Rest turn close st. backw. flex. (str. pos. in forw. flex.)
3. Wg. wlk. st. *a* Kn. flex.
4. Reach st. 2 A. fling.
5. Stoop fall. pos. (Pupil places his hands on the chairs each side of him).
6. Rest wlk. st. *b* T. sidew. flex.
7. 90° upw. jump.
8. St. 2 A. circ.

VII.

1. { Wg. st. alt. toe elev.
   { 2 A. ext. in var. dir.
2. Str. st. T. flex. backw. and forw.
3. Fall hang. H. rot.
4. { Wg. cr. ½ st. Kn. ext. backw. (later also forw.).
5. Bend stoop stride st. alt. A. ext. upw.
6. Wg. sitt. T. backw. flex. (Pupil sits on the desk with his feet under the seat of the chair.)
7. Rest wlk. st. *c* T. sidew. flex.
8. Forw. jump.
9. St. 2 A. elev. sidew. upw.

VIII.

1. { Rest st. 2 Kn. flex.
   { ½ yd. reach st. pos.
2. Rest turn st. backw. flex. (forw. flex. in str. pos.
3. Bal. hang. pos. (Pupil places his hands on the desks each side of him and lifts his feet backward.)
4. Wg. toe st. alt. Kn. flex. upw. in rapid time, followed by Wg. stride st. 2 Kn. flex. to sitt. pos.
5. Str. stoop stride st. 2 A. ext. upw.
6. Rest turn close st. T. sidew. flex.
7. Forw. jump.
8. Yd. wlk. st. *b* 2 A. fling. and change of feet.

IX.

1. { Wg. st. alt. toe- and heel-elev.
   { St. alt. A. fling. upw.
2. Str. stride st. T. flex. backw. and forw.
3. Bend courtesy sitt. 2 A. ext.

4. Forw. ly. pos.
5. Stoop fall. foot placing forw. and backw. (compare VI. 5).
6. Bend turn wlk. st. *a* 2 A. ext. upw.
7. Sidew. jump.
8. St. 2 A. elev. sidew. and heel-elev.

X.

1. { Rest stride st. 2 Kn. flex.
   { Yd. st. 2 A. fling.
2. Rest turn wlk. st. *a* backw. flex. (str. pos. in forw. flex.).
3. Fall hang. alt. L. elev.
4. Bend toe st. 2 A. ext. and 2 Kn. flex.
5. ¼ str. yd. stoop stride st. pos.
6. Rest stride st. T. rot. (also T. sidew. flex.).
7. Sidew. jump.
8. Yd. *d* st. 2 A. elev. and heel-elev.

XI.

1. { Wg. wlk. st. *b* 2 Kn. flex. (also in 6 counts).
   { 2 A. ext. sidew. and backw.
2. Str. stride st. T. flex. backw. forw. downw.
3. Bal. cr. hang. pos. (Compare VIII. 3; knees are lifted forward as in cr. hang. pos.)
4. Rest st. L. clev.
5. Reach stoop stride st. 2 A. fling.
6. Str. stride st. T. rot.
7. With A. fling. upw. jump.
8. St. 2 A. elev. forw. upw. and heel-elev.

XII.

1. { Wg. wlk. st. *a* Kn. flex.
   { Yd. st. 2 A. fling. forw.
2. See XI. 2.
3. Fall hang. alt. L. elev.
4. Rest close st. march on tip-toe.
5. Bend forw. ly. 2 A. ext. sidew.
6. Rest turn wlk. st. *a* T. sidew. flex.
7. See XI. 7. (Also repetition of other forms.)
8. St. 2 A. elev. sidew. upw. and heel-elev.

XIII.

1. { Str. wlk. st. *a* heel-elev.
   { Wg. st. H. rot. with flex.
2. Rest turn stride st. backw. flex.
3. Bal. cr. hang. pos.
4. Wg. fallout st. *a* pos.
5. Yd. forw. ly. 2 A. fling.
6. Bend turn stride st. 2 A. ext. upw.
7. One step's start forw. jump.
8. Yd. *d* toe st. 2 A. elev. and 2 Kn. flex.

XIV.

1. { Str. st. 2 Kn. flex.
   { 2 A. ext. in var. dir.
2. Str. st. T. flex. backw. and forw. downw.
3. Str. close st. march on tip-toe.
4. Rest forw. ly. T. backw. flex.
5. Stoop fall alt. L. elev. (Compare VI. 5.)
6. Wg. fallout st. *a* T. rot.
7. See XIII. 7. (Also over rope or other obstacle.)
8. Yd. turn close st. 2 A. fling.

XV.

1. { Str. wlk. st. *b* heel-elev.
   { Rest stride st. T. rot.
2. See XIV. 2.
3. Fall hang. 2 A. flex.
4. Bend stride courtesy sitt. 2 A. ext.
5. Bend turn fallout st. *a* 2 A. ext. upw.
6. Stoop fall. alt. L. elev.
7. Str. wlk. st. *b* sidew. flex.
8. Two (three) steps' start forw. jump.
9. Yd. turn st. 2 A. circ.

XVI.

1. { Str. wlk. st. *b* 2 Kn. flex.
   { Yd. st. 2 A. fling. forw.
2. Str. turn close st. backw. flex.
3. Fall hang. 2 A. flex.
4. Wg. fallout st. *a* heel-elev.
5. Bend stoop stride st. 2 A. ext. to yd. *d* pos.

APPENDIX

6. Rest sitt. backw. flex. (Compare VII. 6.)
7. Half str. fallout st. *a* pos.
8. See XV. 8. (Also over rope.)
9. St. 2 A. elev. and 2 Kn. flex.

XVII.

1. { Str. stride st. 2 Kn. flex.
   { 2 A. ext. in var. dir.
2. Str. wlk. st. *b* T. flex. backw. forw. downw.
3. Fall hang. 2 A. flex.
4. Wg. fallout st. *b* pos.
5. Yd. *d* stoop stride st. 2 A. elev.
6. Rest sitt. backw. flex.
7. Str. wlk. st. *b* T. rot.
8. Jump. from chair, half start (or other forms of jump.).
9. Yd. fallout st. *a* 2 A. fling.

XVIII.

1. { Str. wlk. st. *a* heel-elev.
   { Bend toe st. 2 A. ext. and 2 Kn. flex.
2. Str. turn st. T. backw. flex.
3. Fall hang. 2 A. flex.
4. Str. cr. half st. pos.
5. Bend forw. ly. 2 A. ext. to yd. *d* pos.
6. Bal. cr. hang. alt. Kn. ext. forw. (Compare XI. 3.)
7. Rest turn stride st. sidew. flex.
8. See XVII. 8.
9. Close st. 2 A. elev. and T. rot.

XIX.

1. Str. turn wlk. st. *a* pos. in two counts.
2. Bend arch stride st. 2 A. ext.
3. Wg. fallout st. *c* pos.
4. Stoop stride st. 2 A. swim. ("*One! Two three!*")
5. Bal. cr. hang. alt. Kn. ext. forw
6. Str. stride st. T. sidew. flex.
7. Twice upw. jump.
8. Yd. turn stride st. 2 A. fling.

XX.

1. Bend courtesy sitt. 2 A. ext. sidew. and upw.
2. Str. turn wlk. st. *a* backw. flex.
3. Stoop fall. 2 A. flex.
4. Str. cr. half st. F. flex. and ext.
5. Reach fallout st. *b* 2 A. fling.
6. Str. foot gr. sitt. backw. flex.
7. Str. turn stride st. sidew. flex.
8. Jump. from chair, whole start. (Repetition of other forms.)
9. Yd. *d* turn st. 2 A. elev.

XXI.

1. { Wg. toe st. 2 Kn. flex. to sitt. pos. (quickly, "*one! Two!*").
   { 2 A. ext. in var. dir.
2. Str. turn stride st. backw. flex.
3. Stoop fall. 2 A. flex.
4. Wg. fallout st. *c* heel-elev.
5. Bend forw. ly. alt. A. ext. upw.
6. See XX. 6.
7. Str. wlk. st. *c* T. rot.
8. 90° twice upw. jump.
9. Yd. fallout st. *b* 2 A. fling.

XXII.

1. { Wg. fallout st. *a* heel-elev.
   { Half str. wlk. st. *c* T. sidew. flex.
2. Str. turn wlk. st. *b* backw. flex.
3. Stoop fall. 2 A. flex. and L. elev.
4. Str. cr. half st. Kn. ext. backw.
5. Yd. turn fallout st. *a* 2 A. elev.
6. Str. stride st. quick. T. rot.
7. Jump. from chair 90°, half start.
8. Yd. turn stride st. 2 A. circ.

XXIII.

1. Half str. fallout st. *a* pos.
2. Str. arch wlk. st. *b* 2 A. ext.
3. Stoop fall. 2 A. flex. and L. elev.
4. Str. cr. half st. Kn. ext. forw.
5. Bend forw. ly. 2 A. ext. upw
6. Str. wlk. st. *c* rev. T. rot.
7. { Twice sidew. jump.
   { Vault. between pommels (over the chair, one hand on the desk, the other on the back rest of the chair).
8. Yd. *d* turn stride st. 2 A. elev.

XXIV.

1. Wg. fallout st. c heel-elev.
2. Str. turn arch wlk. st. a 2 A. ext.
3. Bend fall hang. alt. L. elev
4. Str. st. L. elev.
5. Reach turn fallout st. a 2 A. fling.
6. Rest stride st. quick T. sidew. flex.
7. { Vault. betw. pommels.
   { 180° upw. jump.
8. Yd. arch st. 2 A. fling.

XXV.

1. Yd. d toe st. 2 A. elev. and 2 Kn. flex.
2. Str. turn arch stride st. 2 A. ext.
3. See XXIV. 3.
4. Str. st. L. elev.
5. Str. fallout st. c 2 A. ext. upw.
6. Str. turn wlk. st. a sidew. flex.
7. { Vault. betw. pommels.
   { 180° upw. jump.
8. Yd. arch st. 2 A. fling.

SET No. 3.

Twenty-five tables of exercises for a class of boys, fifteen to eighteen years of age, who have had previous training. Apparatus: horizontal bar, stall-bars, and benches. Lessons last about thirty minutes a day.

(If used for girls, the tables must be changed a little, making certain exercises milder, and the progression slower.)

I.

1. { Feet close and open.
   { H. flex. backw.
   { St. heel-elev.
   { 2 A. ext. upw. and sidew.
2. Wg. st. T. backw. flex.
3. { Over gr. }
   { Und. gr. } hang. pos.
   { Und. }
4. Wg. st. 2 Kn. flex.
5. Half str. yd., etc., pos.
6. Marching.
7. Close st. sidew. flex.
8. Prep. to jump.
9. Yd. a st. 2 A. fling.

II.

1. { Close st. heel-elev.
   { 2 A. ext. forw.
   { Str. stride st. pos.
   { Prep. to jump.
2. Rest st. backw. flex.
3. Climbing the stall-bars.
4. Wg. courtesy sitt. H. rot.
5. Yd. st. 2 A. fling. forw.
6. Marching and running.
7. Stoop fall. H. rot.
8. Rest close st. sidew. flex.
9. Upw. jump.
10. St. 2 A. elev.

III.

1. { St. 2 A. clev.
   { Wg. stride st. heel-elev.
   { Half str. yd., etc., pos.
2. Str. stride st. T. flex. backw. forw. downw.
3. Fall hang. pos.; later 2 A. flex.
4. Yd. toe st. 2 Kn. flex. to sitt. pos.
5. Reach st. 2 A. fling.
6. March. and run.
7. Stoop fall. 2 F. mov.
8. { Rest close st. T. rot., alt. with
   { Rest stride st. sidew. flex.
9. 90° upw. jump.
10. Yd. d st. 2 A. elev.

APPENDIX

IV.

1. { Yd. st. 2 A. fling.
   { Wg. wlk. st. *a* heel-elev.
   { 2 A. ext. forw. upw. and sidew.
2. Str. st. backw. flex.
3. Fall hang. 2 A. flex.
4. Wg. (str.) wlk. st. *a* Kn. flex.
5. Half yd. str. stoop st. ch. of A.
6. March. and run.
7. Str. ly. 2 L. elev. (90°).
8. { Half str. wlk. st. *c* sidew. flex.,
   {   alt. with
   { Str. close st. T. rot.
9. Forw. jump.
10. St. 2 A. elev. forw. upw.

V.

1. { Yd. *d* st. 2 A. elev.
   { Wg. wlk. st. *b* heel-elev.
   { 2 A. ext. sidew. and backw.
2. Gr. arch st. pos.
3. Over gr. hang. osc. trav.
4. Wg. st. L. elev.
5. Reach stoop stride st. 2 A. fling.
6. March. and run.
7. Str. ly. 2 L. elev. (45°).
8. { Bend turn st. 2 A. ext. upw., alt. w.
   { Str. st. sidew. flex.
9. Sidew. jump.
10. Yd. st. 2 A. circ.

VI.

1. { Yd. st. 2 A. circ.
   { Wg. wlk. st. *c* heel-elev.
   { Alt. A. fling. upw.
2. Str. sup. st. backw. flex. over bar.
3. Over gr. hang. osc. trav.
4. Bend courtesy sitt. 2 A. ext.
5. Yd. forw. ly. 2 A. fling.
6. March. and run.
7. Wg. Kn. st. backw. flex.
8. { Str. wlk. st. *a* T. rot., alt. with.
   { Rest turn close st. sidew. flex.
9. 2 A. fling. upw. jump.
10. St. 2 A. elev. sidew. upw.

VII.

1. { St. 2 A. elev. forw. upw.
   { Wg. fallout st. *a* pos.
   { Yd. *a* stoop st. 2 A. fling.
2. Str. wlk. st. *b* backw. flex.
3. Arch hang. and fall hang. pos.
4. Rest st. L. elev.
5. Wg. (rest) forw. ly. T. flex.
6. March. and run.
7. Wg. foot. gr. sitt. backw. flex.
8. { Str. turn wlk. st. *a* 2 A. ext.
   {   upw., alt. with
   { Rest turn wlk. st. *a* sidew. flex.
9. A. and L. fling. upw. jump.
10. St. 2 A. elev. and heel-elev.

VIII.

1. { St. 2 A. and heel-elev.
   { Half str. fallout st. *a* pos.
   { St. alt. toe-elev.
2. Gr. arch st. heel-elev.
3. Fall hang. 2 A. flex., distance between hands increasing.
4. Yd. close toe st. slow march.
5. Bend toe sup. fallout st. *b* 2 A. ext. upw.
6. March. and run.
7. Hor. stoop fall. H. rot.
8. (Wg.) Rest foot side gr. half st. sidew. flex.
9. One step's start. forw. jump.
10. Yd. *d* st. 2 A. and heel-elev.

IX.

1. { Yd. st. 2 A. fling.
   { Wg. fallout st. *b* pos.
   { Bend stoop st. alt. A. ext. upw.
2. Bend arch sup. st. 2 A. ext.
3. Und. hang. osc. trav.
4. Bend toe st. 2 A. ext. and 2 Kn. flex.
5. Str. turn fallout st. *a* 2 A. ext.
6. March. and run.
7. Bend fall foot gr. sitt. 2 A. ext. upw.
8. { Str. stride st. T. rot., alt. with
   { Str. stride st. sidew. flex.
9. Jump. over rope, one step's start.
10. St. 2 A. elev. forw. upw. and heel-elev.

X.

1. { Yd. *d* st. 2 A.- and heel-elev.
   { Wg. wlk. st. *b* 2 Kn. flex.
   { Half str. close st. sidew. flex.
2. Gr. arch stride st. heel-elev.
3. Arch hang. 2 A. flex.
4. Rest toe st. slow march.
5. Bend stoop st. 2 A. ext. to yd. *d*.
6. March. and run.
7. Reach fall. foot gr. sitt. 2 A. fling.
8. Wg. fallout st. *a* T. rot.
9. { Reach gr. st. introd. to vault.
   { Prep. to jump, "*Start!*"
10. Yd. fallout st. *b* 2 A. fling.

XI.

1. { Yd. st. 2 A. fling.
   { Wg. wlk. st. *a* Kn. flex.
   { Str. close st. T. rot.
2. Gr. arch st. Kn. upw. flex.
3. "Incl. rope climb." (hor. incl. bar, put low).
4. Bend stride courtesy sitt. 2 A. ext.
5. Str. stoop stride st. 2 A. ext. to yd. *e*.
6. March. and run.
7. Rest foot gr. sitt. backw. flex.
8. Rest turn close st. backw. flex.
9. { Half yd. gr. st. introd. to vault.
   { Twice upw. jump.
10. St. 2 A. elev. sidew. upw. and heel-elev.

XII.

1. { Yd. fallout st. *b* 2 A. fling.
   { Wg. stride st. 2 Kn. flex.
   { Bend turn st. 2 A. ext. upw. and sidew.
2. Str. arch st. 2 A. ext. upw.
3. See XI. 3.
4. Rest st. Kn. flex. upw.
5. Yd. stoop stride st. 2 A. elev.
6. March. and run.
7. Stoop fall. 2 A. flex.
8. Str. wlk. st. *b* T. rot. or sidew. flex.
9. { Introd. to vault., one step's start.
   { 90° twice upw. jump.
10. Yd. *d* toe st. 2 A. elev. and 2 Kn. flex.

XIII.

1. { See XII. 10.
   { H. sidew. flex.
   { Rest turn close st. sidew. flex.
2. Fall hang. to gr. arch st. pos.
3. Over gr. hang. osc. trav.
4. Wg. hor. half st. pos.
5. Bend foot gr. fallout st. *b* 2 A. ext.
6. March. and run.
7. Rest Kn. st. backw. flex.
8. Half str. rev. turn fallout st. *a* pos.
9. Wg. st. alt. toe-elev.
10. Cr. hang. pos. (stall-bars).
11. { Introd. to vault., 3 steps' start.
    { Jump. from bench.
12. Yd. walking 2 A. fling.

XIV.

1. { 2 A. elev. forw. upw. and heel-elev.
   { Str. st. 2 Kn. flex.
   { Half str. wlk. st. *c* sidew. flex.
2. Str. arch wlk. st. *b* 2 A. ext.
3. "Incl. rope climb.," feet first (bar horizontal).
4. Str. cr. half st. F. flex. and ext.
5. Yd. turn fallout st. *a* 2 A. elev.
6. March. and run.
7. Wg. foot gr. half st. backw. flex.
8. Str. turn wlk. st. *a* 2 A. ext. and ch. of F.
9. Fallout st. *c* pos.
10. Cr. hang. pos. (hor. bar).
11. { Vault. as in XIII. 11, bar put higher.
    { 90° jump. from bench.
12. Fallout st. *b* 2 A. elev. forw. upw.

XV.

1. { 2 A. elev. sidew. upw. and heel-elev.
   { Str. wlk. st. *b* 2 Kn. flex.
   { Bend turn stride st. 2 A. ext.
2. Gr. arch st. Kn. flex. and ext.
3. Und. hang. osc. trav.
4. Str. toe st. slow. march.
5. Bend forw. ly. 2 A. ext. to yd. *d*.
6. March. and run.
7. Str. Kn. st. backw. flex.

APPENDIX

8. Str. turn wlk. st. *a* backw. flex.
9. Reach gr. st. toe- and heel-elev.
10. Cr. hang. alt. Kn. ext.
11. { Vault. over benches, face down.
    Jump. over rope (length), "whole start."
12. Yd. turn st. 2 A. fling.

XVI.

1. { Yd. turn st. 2 A. fling.
     Str. stride st. 2 Kn. flex.
     2 A. ext. in var. dir.
2. Gr. arch st. L. elev.
3. See XIV. 3, bar inclined.
4. Rest cr. half st. Kn. ext. backw.
5. Str. fallout st. *b* forw. flex.
6. Hor. stoop fall. 2 A. flex.
7. March. and run.
8. Wg. side fall. pos.
9. Str. wlk. st. *c* T. rot.
10. Wg. st. slow 2 Kn. flex. to sitt. pos.
11. Cr. hang. alt. Kn. ext.
12. { Somersault over bar.
     Jump. over rope (height), "whole start."
13. Yd. *d* fallout st. *b* 2 A. elev.

XVII.

1. { Yd. *d* fallout st. *b* 2 A. elev.
     Str. wlk. st. *a* Kn. flex.
     Half str. yd. stoop st. ch. of A.
2. Gr. arch st. L. elev.
3. Und. gr. hang. 2 A. flex.
4. Str. cr. half st. Kn. ext. backw.
5. Str. foot gr. fallout st. *b* Kn. flex.
6. Rest foot gr. half st. backw. flex.
7. March. and run.
8. Str. side sup. wlk. st. *b* sidew. flex.
9. Wg. side fall. L. elev.
10. Str. wlk. st. *c* heel-elev.
11. Cr. hang. 2 Kn. ext.
12. { Vault. bar, face down; start without run., bar quite low.
     Jump. over rope; name foot for start.
13. Yd. turn st. 2 A. circ.

XVIII.

1. { 2 A. elev. forw. upw. and heel-elev.
     Yd. toe st. 2 Kn. flex. to sitt. pos. (quickly).
     Bend stoop st. 2 A. ext. sidew. and upw.
2. Fall hang. to gr. arch st. (bar gradually lowered).
3. Und. hang. 2 A. flex.
4. Rest st. L. elev.
5. Str. foot gr. fallout st. *b* 2 A. fling.
6. Incl. stoop fall. 2 A. flex.
7. Str. foot side gr. half st. sidew. flex.
8. March. and run.
9. Str. turn wlk. st. *a* backw. flex.
10. Instep gr. st. Kn. flex.
11. Cr. hang. 2 Kn. ext.
12. { Vault. bar, face down, 3 steps' start.
     Star gr. half st. jump. from stallbars.
13. Yd. turn stride st. 2 A. fling.

XIX.

1. { Fallout st. *b* 2 A. elev. forw. upw.
     Yd. courtesy sitt. H. rot.
     Str. stride st. sidew. flex.
2. Gr. arch st. L.- and heel-elev.
3. Over gr. hang. 2 A. flex.
4. Str. cr. half st. Kn. ext. forw.
5. Bend forw. ly. 2 A. ext. upw.
6. March. and run.
7. Str. Kn. half st. backw. flex.
8. Str. side gr. st. L. elev.
9. Str. turn st. backw. flex.
10. Wg. fallout st. *a* heel-elev.
11. Cr. hang. 2 Kn. ext. (hor. bar).
12. { Vault. bar, face up, whole start.
     90° jump. over rope.
13. See XVIII. 13.

XX.

1. { H. rot. with flex.
     Bend courtesy sitt. 2 A. ext.
     Yd. st. 2 A. fling. forw.
2. Gr. arch stride st. heel-elev. (deep flex.).

3. "Incl. rope climb.," feet first (incl. bar).
4. Str. st. L. elev.
5. Str. foot gr. fallout st. *b* forw. flex.
6. Fall Kn. st. 2 A. ext. upw.
7. March. and run.
8. Str. turn wlk. st. *a* backw. flex.
9. Str. side sup. st. sidew. flex. and L. elev.
10. Str. wlk. st. *b* T. forw. flex. (foot against bar).
11. Cr. hang. 2 L. elev.
12. { Vault. bar, face up, run for start. Sidew. jump. over rope.
13. Yd. *d* turn st. 2 A. elev.

XXI.

1. { Yd. *d* turn st. 2 A. elev. Bend toe st. 2 A. ext. and 2 Kn. flex. Wg. fallout st. *a* T. rot.
2. Gr. arch st. heel-elev. (deep flex.).
3. Und. gr. hang. trav.
4. Rest. cr. half st. Kn. abd.
5. Yd. foot gr. fallout st. *b* 2 A. elev.
6. Fall Kn. half st. 2 A. ext. upw.
7. March. and run.
8. Str. turn stride st. backw. flex.
9. Half str. gr. side fall. L. elev.
10. Half yd. gr. half st. Kn. flex.
11. Cr. hang. 2 L. elev.
12. { Vault. bar, face up or down; increasing height. Somersault, bar gradually lowered. Jump. from stall-bars, increasing height.
13. Yd. turn stride st. 2 A. circ.

XXII.

1. { Yd. turn st. 2 A. fling. Bend courtesy stride sitt. 2 A. ext. Rest turn close st. backw. flex.
2. Gr. arch st. 2 A. flex.
3. Und. gr. hang. trav.
4. Rest cr. half st. Kn. ext. forw.
5. Str. forw. ly. forw. flex.
6. March. and run.
7. Reach fall Kn. st. 2 A. fling.
8. Str. fallout st. *a* T. rot.

9. Str. stride st. quick. sidew. flex.
10. Half str. yd. gr. hor. half st. Kn. flex.
11. Cr. hang. 2 L. elev. (hor. bar).
12. See XXI. 12.
13. Toe-sup. fallout st. *b* 2 A. elev. forw. upw.

XXIII.

1. { Yd. *d* fallout st. *b* 2 A. elev. Feet close and open rhythmically (twelve times). 2 A. ext. in var. dir.
2. Gr. arch stride st. hand to hand mov.
3. Und. hang. trav.
4. Str. hor. half st. pos.
5. Str. fallout st. *c* 2 A. elev.
6. March. and run.
7. Reach fall Kn. half st. 2 A. fling.
8. Str. wlk. st. *c* rev. T. rot.
9. Introd. to wheeling.
10. Wg. fallout st. *c* heel-elev.
11. Cr. hang. 2 L. abd.
12. { Vault. bar, 180° turn. Jump. over rope, whole start (high and far).
13. Yd. *d* turn stride st. 2 A. elev.

XXIV.

1. { Yd. *d* toe st. 2 A. elev. and Kn. flex. Wg. fallout st. *a* rev. T. rot. Str. stoop stride st. 2 A. ext. to yd. *e*.
2. Gr. arch stride st. hand under hand mov.
3. Und. hang. trav. on incl. bar.
4. Reach half st. Kn. flex., also Rest toe st. slow march.
5. Rest foot gr. fallout st. *b* forw. flex.
6. Bend fall foot gr. half st. 2 A. ext.
7. March. and run.
8. Half str. side fall. L. elev.
9. Wheeling.
10. Heel gr. half st. T. forw. flex.
11. Cr. hang. 2 L. abd.
12. See XXIII. 12.
13. Yd. arch st. 2 A. fling.

XXV.

1. {
   2 A. elev. sidew. upw. and 2 Kn. flex.
   St. alt. toe-elev. rhythmically.
   Half str. rev. turn fallout st. *a* pos.
   }
2. Gr. arch stride st. 2 A. flex. (T. flex. deep).
3. Und. hang. trav. with alt. elev.
4. Reach half st. Kn. flex.; also Str. toe st. slow. march.
5. Str. fallout st. c forw. flex.
6. March. and run.
7. Str. foot gr. half st. backw. flex.
8. Str. stride st. quick T. rot.
9. Wheeling.
10. Yd. courtesy sitt. 2 A. elev.
11. Cr. hang. 2 L. abd. (hor. bar).
12. {
    Repetition of previous forms of vault.
    Jump. over rope (height); run for start.
    }
13. See XXIV. 13.

SET No. 4.

Twenty-five tables of exercises for a class of men twenty to twenty-five years old, who have had previous training.

The gymnasium has all the apparatus belonging to the system.

Lessons last from thirty to forty-five minutes a day.

Marching is put in wherever the teacher thinks best.

I.

1. {
   St. 2 A. elev. and heel-elev.
   2 A. ext. upw. and sidew., etc.
   }
2. Str. st. (stride st.) T. flex. backw. forw. downw.
3. { Over gr. hang. or Und. hang. } osc. trav.
4. Wg. st. 2 Kn. flex.
5. Reach st. 2 A. fling.
6. Half str. close st. sidew. flex.
7. Prep. to jump.
8. Yd. st. 2 A. fling.

II.

1. {
   Yd. *d* st. 2 A. elev.
   Wg. wlk. st. *a* pos. in series.
   2 A. ext. in var. dir.
   Prep. to jump.
   }
2. { Gr. arch st. heel-elev., alt. with Str. st. back. flex. over bar. }
3. { Under gr. hang. Und. hang. } 2 A. flex.
4. Str. st. (stride st.) 2 Kn. flex.
5. Yd. st. 2 A. fling. forw.
6. Str. gr. st. Kn. flex. and ext.
7. Rest close st. sidew. flex.
8. Str. close st. T. rot.
9. { Upw. jump.; later 90° upw. jump. and A. fling. upw. jump. }
10. { 2 A. elev. forw. upw., followed by Yd. st. 2 A. fling. }

III.

1. {
   St. 2 A. elev. forw. upw.
   Wg. wlk. st. *a* heel-elev. in series.
   Half str. yd., etc., pos.
   }
2. { Gr. arch stride st. heel-elev., alternating with Str. arch stride st. 2 A. ext }
3. Over gr. hang. 2 A. flex.
4. Wg. st. Kn. flex. upw. and ext. backw.
5. Reach stoop st. 2 A. fling.
6. Str. ly. L. elev.
7. { Bend turn stride st. 2 A. ext. upw., alt. with Wg. turn close st. backw. flex. }
8. St. alt. toe-elev.
9. A. and L. fling. upw. jump. (also see II. 9).
10. St. 2 A. circ.

IV.

1. { St. 2 A. circ.
   Wg. fallout st. *a* (later *b*) pos.
   2 A. ext. in var. dir.
2. Str. arch wlk. st. *b* 2 A. ext. (later also with change of feet).
3. Fall hang. and arch hang. pos.; later fall hang. 2 A. flex.
4. Wg. st. L. elev.
5. Rest forw. ly. T. flex.
6. Stoop fall pos., later 2 A. flex.
7. { Str. st. sidew. flex., alt. with Rest stride st. T. rot.
8. St. alt. toe- and heel-elev.
9. { Reach gr. and half yd. gr. st. introd. to vault.
   Forw. jump.; also sidew. jump.
10. St. 2 A. and heel-elev.

V.

1. { St. 2 A. and heel-elev.
   Wg. wlk. st. *b* heel-elev. in series.
   H. flex. sidew., etc.
   Reach st. 2 A. fling.
2. Fall hang. to gr. arch st. pos.
3. Zigzag (or hor.) serpentine in hor. ladder.
4. Wg. cr. half st. Kn. ext. forw.
5. Str. stoop st. 2 A. ext. to yd. *e*.
6. Wg. Kn. st. backw. flex.
7. Str. wlk. st. *b* T. rot.
8. Fallout st. *c* pos.
9. { Introd. to vault. (bar grad. raised).
   Forw. jump. 1 (3) step's start.
10. Yd. *d* st. 2 A. and heel-elev.

VI.

1 { St. 2 A. and heel-elev.
  Wg. wlk. st. *b* 2 Kn. flex. in series.
  Half str. close st. sidew. flex.
2 Rest arch st. heel-elev. (living supports).
3. Climb. on rope-ladder.
4. Wg. cr. half st. Kn. abd.
5. Yd. stoop st. 2 A. elev.
6. Wg. (Rest) foot gr. sitt. backw. flex.
7. Wg. foot side gr. half st. sidew. flex.
8. Rest turn close st. backw. flex.
9. Str. wlk. st. *c* heel-elev.

10. Cr. hang. pos. (stall-bars).
11. { Vault along double-bar, face up.
    90° forw. jump. 1 (3) step's start.
12. { Yd. fallout st. *b* 2 A. fling., followed by
    St. 2 A. elev. forw. upw.

VII.

1. { Yd. st. 2 A. fling.
   Wg. wlk. st. *a* Kn. flex.
   Str. st. T. rot.
2. Gr. arch st. Kn. flex. upw.
3. Zigzag serpentine in vertical ladder.
4. Bend toe st. 2 A. ext. and 2 Kn. flex.
5. Bend forw. ly. 2 A. ext. to yd. *d*.
6. Str. foot gr. sitt. 2 A. ext.
7. Rest turn stride st. sidew. flex.
8. Toe sup. wlk. st. *b* Kn. flex.
9. Cr. hang. pos. (hor. bar).
10. { Vault. as in VI., bars closer together (the lower one raised).
    Jump. from bench.
11. St. 2 A. elev. forw. upw. and heel-elev.

VIII.

1. { St. 2 A. elev. forw. upw. and heel-elev.
   H. rot. with flex.
   Rest wlk. st. *a* Kn. flex.
2. Gr. arch st. Kn. flex. and ext.
3. Incl. rope climb.
4. Wg. toe st. alt. Kn. flex. upw.
5. Yd. turn fallout st. *a* 2 A. elev.
6. Reach foot gr. sitt. 2 A. fling.
7. Half str. rev. turn fallout st. *a* pos.
8. Str. wlk. st. *b* sidew. flex.
9. Wg. st. slow. 2 Kn. flex.
10. Cr. hang. alt. Kn. ext.
11. { Somersault over the bar.
    Jump. over rope, whole start.
12. Yd. *d* toe st. 2 A. elev. and 2 Kn. flex.

IX.

1. { St. 2 A. elev. forw. upw. and heel-elev. [slowly.
   Wg. st. 2 Kn. flex. to sitt. pos.,
   Str. st. sidew. flex.
2. Gr. arch st. L. elev.

APPENDIX

3. Vertical serpentine (ladder).
4. Rest st. Kn. flex. upw. and ext. backw.
5. Reach turn fallout st. *a* 2 A. fling.
6. Rest foot gr. sitt. backw. flex. to hor. pos.
7. Str. turn wlk. st. *a* 2 A. ext. and ch. of F.
8. Wg. side fall. pos.
9. Rest wlk. st. *a* Kn. flex.
10. Incl. rope climb.
11. { Vault. betw. pommels (saddles).
    { Jump. over rope, 1 (3) step's start.
12. Yd. walking 2 A. fling.

X.

1. { Yd. *d* st. 2 A. and heel-elev.
   { Feet close and open, rhythmically.
   { Rest stride st. T. rot.
   { Yd. st. 2 A. fling. forw.
2. Gr. arch st. L. and heel-elev.
3. Und. hang. L. swing. over bar (double bar).
4. Rest st. L. elev.
5. Rest foot gr. fallout st. *b* Kn. flex.
6. Rest Kn. (half) st. backw. flex.
7. Rest foot side gr. half st. sidew. flex.
8. Str. turn wlk. st. *a* backw. flex.
9. Wg. st. slow 2 Kn. flex. to sitt. pos.
10. Arch hang. 2 A. flex.
11. { Vault. double bar, face down or up.
    { 90° jump. over rope, 1 (3) step's start.
12. { Yd. turn st. 2 A. fling.
    { Fallout st. *b* 2 A. elev. forw. upw.

XI.

1. { Yd. turn st. 2 A. fling.
   { Wg. fallout st. *a* heel-elev.
   { Rest turn stride st. backw. flex.
   { 2 A. ext. in var. dir.
2. Gr. arch st. 2 A. flex.
3. Vert. rope climb.
4. Str. crosswise st. 2 Kn. flex.
5. Str. foot gr. fallout st. *b* forw. flex.
6. Rest foot gr. half st. backw. flex.
7. Str. side sup. wlk. st. *b* sidew. flex.
8. Str. wlk. st. *c* T. rot.

9. Instep gr. half st. Kn. flex.
10. Cr. hang. alt. Kn. ext.
11. { Vault. bar, face down or up.
    { Jump. rope, name foot for start.
12. Yd. *d* fallout st. *b* 2 A. elev.

XII.

1. { St. 2 A. elev. forw. upw. and heel-elev.
   { Yd. toe st. 2 Kn. flex. to sitt. pos., quickly.
   { Yd. stoop st. 2 A. fling.
   { Str. stride st. T. rot.
2. Gr. arch stride st. hand to hand mov.
3. Vert. pole climb.
4. Str. cr. half st. Kn. ext. backw.
5. Str. turn fallout st. *a* forw. flex.
6. Hor. (incl.) stoop fall. 2 A. flex.
7. Str. turn st. backw. flex.
8. Str. foot side gr. half st. sidew. flex.
9. Heel gr. half st. T. forw. flex.
10. Cr. hang. 2 Kn. ext. (stall-bars).
11. { Vault. over horse, face down or up.
    { 90° jump. from bench.
12. { Yd. turn stride st. 2 A. fling.
    { St. 2 A. elev. sidew. upw. and 2 Kn. flex.

XIII.

1. { Fallout st. *b* 2 A. elev. forw. upw.
   { Wg. stride st. 2 Kn. flex. to sitt. pos., slowly.
   { Reach stoop st. 2 A. fling. upw.
   { Str. stride st. sidew. flex.
2. Oblique gr. arch st. heel-elev.
3. Vert. rope climb., from rope to rope.
4. Wg. hor. half st. pos.
5. Yd. *b* st. 2 A. fling. to yd. *e*.
6. Str. Kn. (half) st. backw. flex.
7. Wg. side fall. L. elev.
8. Str. wlk. st. *c* T. rot.
9. Wg. fallout st. *a* heel-elev.
10. Cr. hang. 2 Kn. ext. (hor. bar).
11. { Vault. over bar, 180° turning.
    { Twice upw. jump., etc.
12. { Close st. 2 A. elev. w. T. rot.
    { Yd. st. 2 A. fling.

XIV.

1. { St. 2 A. elev. sidew. upw. and 2 Kn. flex.
    Wg. fallout st. *a* (*b*) pos. backw.
    Str. stoop stride st. 2 A. ext. upw.
    Str. wlk. st. *c* sidew. flex. }
2. Fall hang. to gr. arch st. pos. (bar gradually lowered).
3. Pole climb., from pole to pole.
4. Str. cr. half st. Kn. ext. forw.
5. Rest foot gr. fallout st. *b* forw. flex.
6. Bend fall. Kn. (half) st. 2 A. ext.
7. Rest fallout st. *a* T. rot.
8. Str. turn close st. backw. flex.
9. Reach half st. Kn. flex. to 90°.
10. Cr. hang. 2 L. elev. (stall-bars).
11. { Vault. over horse (box) lengthwise.
     Jump. from stall-bars. }
12. Yd. *d* turn st. 2 A. elev.

XV.

1. { St. 2 A. elev. forw. upw. & heel-elev.
    Half str. fallout st. *a* pos.
    Bend toe st. 2 A. ext. and 2 Kn. flex. quickly.
    Str. stride st. T. flex. backw. forw. downw. }
2. Oblique gr. arch stride st. heel-elev.
3. Und. gr. hang. trav.
4. Str. st. L. elev.
5. Bend forw. ly. 2 A. ext. upw.
6. Reach fall. Kn. (half) st. 2 A. fling.
7. Str. fallout st. *a* T. rot.
8. Str. turn stride st. sidew. flex.
9. Wg. fallout st. *c* heel-elev.
10. Cr. hang. 2 L. elev. (hor. bar).
11. { Vault. over bar with rope.
     Forw. jump. etc. ("Stárt! Five!") }
12. Yd. turn fallout st. *a* 2 A. fling.

XVI.

1. { Yd. turn stride st. 2 A. fling.
    Rest st. slow 2 Kn. flex.
    Half str. yd., etc., pos.
    Str. stride st. quick. T. rot. }
2. Gr. arch st. L. and heel-elev. (deep T. flex.).
3. Under hang. trav. (hor. bar).
4. Str. cr. half st. Kn. abd.
5. Str. fallout st. *c* 2 A. ext.
6. Bend fall foot gr. half st. 2 A. ext.
7. Str. turn wlk. st. *a* sidew. flex.
8. Str. side gr. half st. L. elev.
9. Str. wlk. st. *b* forw. flex. (foot braced against bar).
10. Cr. hang. 2 L. abd.
11. { Somersault over bar (gradually lowered).
     Vault. over horse lengthwise, etc.
     Free st. jump. as above: "Stárt!" }
12. Toe sup. fallout st. *b* 2 A. elev. forw. upw.

XVII.

1. { Toe sup. fallout st. *b* 2 A. elev. forw. upw.
    Str. st. 2 Kn. flex. slowly: "Stárt!"
    H. rot. w. flex.
    2 A. ext. in var. dir. (series). }
2. Gr. arch stride st. 2 A. flex.
3. Und. hang. trav. on incl. bar.
4. Rest cr. half st. Kn. ext. forw.
5. Str. hor. half st. Kn. flex.
6. Str. foot gr. half st. backw. flex.
7. Half str. gr. side fall. L. elev.
8. Str. turn wlk. st. *b* backw. flex.
9. Half yd. gr. half st. Kn. flex. (to sitt. pos.).
10. Cr. hang. 2 L. abd.
11. { Vault. betw. pommels on horse.
     Vault. with rope, bar high.
     Free st. jump. as above. }
12. Yd. *d* turn stride st. 2 A. elev.

XVIII.

1. { Yd. *d* st. 2 A. & heel-elev. ["Stárt!"
    Str. stride st. 2 Kn. flex. slowly:
    Rest turn stride st. sidew. flex.
    2 A. ext. in series. }
2. Gr. arch stride st. hand under hand mov.
3. Incl. rope climb.
4. Str. cr. half st. Kn. abd.
5. Yd. foot gr. fallout st. *b* 2 A. elev.
6. Reach fall foot gr. half st. 2 A. fling.
7. Yd. stride st. sidew. flex.
8. Str. wlk. st. *c* rev. T. rot.
9. Half str. yd. gr. hor. half st. Kn. flex.

APPENDIX

10. Und. hang. trav. on ropes.
11. { Sidew. vault. over bar.
    { Jump. from stall-bars (high).
12. Turn fallout st. *a* 2 A. elev. forw. upw.

XIX

1. { Yd. *d* toe st. 2 A. elev. and 2 Kn. flex.
   { Bend courtesy sitt. 2 A. ext. sidew. and upw.
   { Rest stride st. quick sidew. flex.
2. Oblique gr. arch st. L. elev.
3. Und. hang. trav. w. alt. elev.
4. Bal. walk on hor. bar.
5. Str. forw. ly. forw. flex.
6. Wg. foot gr. half st. backw. flex. w. Kn. flex.
7. Str. rev. turn fallout st. *a* sidew. flex.
8. Str. stride st. quick T. rot.
9. Str. stride st. slow. 2 Kn. flex. to sitt. pos.
10. Somersault betw. ropes.
11. { Vault. backw. over horse.
    { Jump. rope; height; 90° turning.
12. { Yd. arch st. 2 A. fling.
    { St. 2 A.- and heel-elev.

XX

1. { Yd. arch st. 2 A. fling.
   { Half str. fallout st. *a* ch. of A.
   { Str. close st. quick T. rot.
2. Gr. arch stride st. 2 A. flex. (deep T. flex.).
3. Over gr. hang. trav.
4. Half str. hor. half st. pos.
5. Yd. fallout st. *c* 2 A. elev.
6. Stoop fall. 2 A. flex. and L. elev.
7. Half str. side fall. L. elev.
8. Str. turn arch wlk. st. *a* 2 A. ext.
9. Bend toe st. slow 2 A. ext. and 2 Kn. flex. to sitt. pos.
10. Cr. hang. somersault (und. gr.)
11. { Vault. over upper bar.
    { Jump. over rope, increasing height.
12. Stride st. 2 A. elev. w. T. rot.

XXI

1. { Yd. turn st. 2 A. fling.
   { Str. crosswise st. 2 Kn. flex.
   { Bend turn wlk. st. *a* 2 A. ext. upw.
   { Prep. to jump. ("Stárt !")
2. Oblique gr. arch st. 2 A. flex.
3. Over gr. hang. 2 hand trav.
4. Bal. walk. backw. on hor. bar.
5. Yd. *b* stoop stride st. 2 A. fling. to yd. *e*.
6. Str. foot gr. sitt. backw. flex. (45°).
7. Str. turn stride st. backw. flex.
8. Str. stride st. quick sidew. flex.
9. Str. stride courtesy sitt. 2 A. ext.
10. Cr. hang. somersault. (over gr.).
11. { Vault. over horse, face down, and 180° turn.
    { Run. jump. over rope, increasing distance.
12. Yd. *d* turn fallout st. *a* 2 A. elev.

XXII

1. { Yd. *d* turn st 2 A. elev.
   { Str. wlk. st. *a* Kn. flex.
   { Rest turn close st. sidew. flex.
   { 2 A. ext. in series.
2. Oblique gr. arch st. L. and heel-elev.
3. Incl. rope climb feet first.
4. Str. hor. half st. pos.
5. Str. fallout st. *c* forw. flex.
6. Str. foot gr. sitt. backw. flex. to hor. pos.
7. Introd. to wheeling.
8. Str. turn arch stride st. 2 A. ext.
9. Str. wlk. st. *a* Kn. flex.
10. Over gr. cr. hang. 2 A. flex. (hor. bar).
11. { Vault. over bar, face down or up: incr. height.
    { Forw. jump. over two marks on the floor; whole start (increasing distance).
12. { Yd. arch st. 2 A. circ.
    { St. 2 A. elev. forw. upw. and heel-elev.

XXIII.

1. { Yd. *d* turn stride st. 2 A. elev.
   Half str. fallout st. *a* pos., zigzag forw.
   Half str. yd., etc., pos.
   Str. turn stride st. sidew. flex. }
2. Oblique gr. arch stride st. 2 A. flex.
3. Und. gr. hang. 2 hand trav.
4. Reach half st. Kn. flex.
5. Str. fallout st. *c* 2 A. ext. and ch. of F. backw.
6. Rev. stoop fall. pos. (stall-bars; later, living supports).
7. Str. side sup. st. sidew. flex. and L. elev.
8. Str. turn arch wlk. st. *a* 2 A. ext. and ch. of F.
9. Yd. courtesy sitt. 2 A. elev.
10. Und. hang. trav. on ropes; somersault when ceiling is reached.
11. { Somersault on saddles (first introd. to same).
    Run. high jump. }
12. Yd. arch wlk. st. *b* 2 A. circ.

XXIV.

1. { Close st. 2 A. elev. and T. rot.
   Half str. fallout st. *b* ch. A. and F. forw., one count.
   Rest stride st. slow 2 Kn. flex.
   Str. wlk. st. *c* rev. T. rot. }
2. Gr. arch stride st. 2 hand mov.
3. Rotary climb. on incl. rope.
4. Yd. st. L. elev.
5. Yd. forw. ly. 2 A. elev.
6. Rev. stoop fall. 2 A. flex. (stall-bars).
7. Sidew. hang. pos.
8. Wheeling.
9. Yd. stride st. quick. T. rot.
10. Str. hor. half st. Kn. flex.
11. Cr. und. hang. trav. on hor. bar.
12. { Vault. backw. over horse.
    Somersault on vault. box.
    Run. long jump. over rope. }
13. { Yd. arch wlk. st. *b* 2 A. fling. and ch. of F.
    St. 2 A. elev. forw. upw. and heel-elev. }

XXV.

1. { Stride st. 2 A. elev. and T. rot.
   Half str. fallout st. *a* pos., zigzag backw.
   Str. turn. stride st. backw. flex. }
2. Oblique gr. arch st. L. and heel-elev. w. 2 A. flex.
3. Rotary trav. on bar, ended by cr. hang. somersault.
4. { Str. toe st. L. elev., alt. with Str. toe st. alt. Kn. flex. upw. slowly. }
5. Bend hor. half st. 2 A. ext. upw.
6. Rev. stoop fall. 2 A. flex. (liv. sup.).
7.
8. { See XXIV.
9.
10. Reach half st. Kn. flex.
11. Cr. und. hang. trav. on vert. ropes.
12. { Vault. over upper bar.
    Somersault on vault. box.
    Run. high jump. over rope, etc. }
13. { Str. arch sup. stride st. 2 A. ext. upw.
    St. 2 A. elev. sidew. upw. and 2 Kn. flex. }

COMPLETE INDEX TO THE NOMENCLATURE.

A. = arm (for arm-movements see pp. 88 to 90 and 121 to 137).
abd. = abduction (pp. 67, 102, 216, 217).
add. = adduction (pp. 102, 216, 217).
alt. = alternate (pp. 55, 66, 94, 121).
arch = the trunk bent backward (Fig. 57, p. 72).
arch hang. = the body is suspended by the arms, the toes resting on the floor behind (Fig. 106, p. 106).
B. = back.
backw. = backward.
bal. = balance.
bal. hang. = Fig. 103, p. 103, and Fig. 207, p. 180.
bend. = "arms upward — bend!" (Fig. 80, p. 88.)
betw. = between.
ch. = change.
circ. = circumduction (pp. 127, 206).
climb. = climbing (pp. 106 to 119 inclusive).
close st. = "feet — clóse!" (see p. 42.)
close toe st. = Fig. 31, p. 53.
courtesy st. = the knees are bent to 90° (Fig. 36, p. 56).
courtesy sitt. = the knees are bent to utmost flexion (Fig. 39, p. 58).
cr. = crook = hips (and knees) bent to 90°.
cr. hang. = crook hanging: — Figs. 96, 97, 99, 100, 101.
cr. ½ st. ⎫ crook half standing: — standing on one leg, the other raised for-
cr. hlf. st. ⎭ ward with hip and knee bent to 90° (Fig. 50, p. 65).
crosswise st. = crosswise standing (Fig. 24, p. 48).
dir. = direction.
elev. = elevation.
ext. = extension.
F. = foot.
fall = the trunk inclined backward (Figs. 157 and 158, p. 140).
fall. = falling.
fall hang. = fall hanging: — the body suspended by the arms; the heels rest on the floor in front (Fig. 105, p. 105).
fallout st. = fallout standing: — one foot in front of the other; the distance between the heels equal to three times the length of the foot; the forward knee bent, the backward one straight.

fallout st. *a* = Fig. 25, p. 49.
fallout st. *b* = Fig. 229, p. 205.
fallout st. *c* = Fig. 29, p. 52.
fallout st. *d* = toe sup. fallout st. *b* (Fig. 149, p. 134).
flex. = flexion.
fig. } = flinging (pp. 124 to 126).
fling.
foot gr. = foot grasp: — one or both feet are inserted between the stall-bars or other apparatus (Figs. 26, 27, 160, etc.).
forw. = forward.
forw. ly. = forward lying: — lying face down ("prone lying") (Fig. 142, p. 129).
fund. = fundamental.
gr. = grasp: — the hands or feet grasp some apparatus.
H. = head (for head-movements see pp. 44, 45).
half
hlf. } = half: — only one limb takes the position following this word.
½
hlf. st. (= ½ st.) = standing on one foot.
hor. = horizontal.
hor. hlf. st. = one leg raised backward to horizontal position in a line with the body, which inclines forward; the supporting leg is bent (Fig. 28, p. 52).
hor. stoop fall. = stoop fall. (see below), except that feet and shoulders are on a level (Fig. 169, p. 146).
incl. = inclined.
incl. stoop fall. = stoop fall (see below), except that the feet are higher than the shoulders (Figs. 170 and 171, p. 147).
introd. = introduction.
jump. = jumping (jumping exercises, p. 171, etc.).
K. (Kn.) = knee.
Kn. st. = knee standing = kneeling (Fig. 175, p. 140).
L. = leg (for leg-movements see p. 46, etc., and p. 162, etc.).
(l.) = left.
ly. = lying, the body extended, face up ("back lying").
march. = marching (for marching see p. 39, etc.).
mov. = movement.
osc. = oscillatory (pp. 22, 94, etc.).
pos. = position (pp. 29, 35).
prep. = preparation.
(r.) = right.
reach = arms extended horizontally forward (Fig. 81 *b*, p. 89).
rest. = the hands are locked behind the neck (Fig. 10, p. 43).
rev. = reverse.
rev. rot. = rotation of the trunk in direction of the backward foot (see p. 156, first line from above, etc.).
rev. stoop fall. = "standing on the hands" (Fig. 173, p. 149).

rot. = rotation.
run. = running (p. 41).
sidew. = sideways.
sitt. = sitting.
st. = standing; if preceded by other abbreviations, it means that the parts not mentioned are in fundamental position.
star st. = feet and arms are apart, so that the position resembles a star (Fig. 194 and 205).
std. = stride (see below).
stoop (stp.) = the trunk bent forward (Fig. 58, p. 72).
stoop fall. = stoop falling: — hands and feet are on the floor (Fig. 168, p. 145).
stoop hang. = hanging with feet up, head down (Figs. 93, 95, 113, 114).
str. = stretch: — the arms extended upward (Fig. 81 a, p. 89).
stride (std.) = the feet apart and on a level (Fig. 23, p. 48).
sup. = support: — the trunk resting on a firm support.
swim. = swimming (p. 227).
swing. = swinging (p. 101).
T. = trunk.
toe st. = the heels raised above the floor (Fig. 30, p. 53).
toe sup. = one foot has the heel raised above the floor (Figs. 40 and 149).
trav. = travelling (p. 94 and following).
turn. = the trunk rotated.
und. gr. hang. = under grasp hanging (Fig. 85, p. 93).
und. hang. = under hanging: — one hand each side the bar, or hanging right under the apparatus (Fig. 98).
upw. = upward.
var. = various.
vault. = vaulting (for vaulting exercises see p. 179, etc.).
w. = with.
wg. = wing (Fig. 15, p. 43).
wlk. st. = walk st.: — one foot is in front of the other; the distance between the heels is equal to twice the length of the foot; both knees are straight.
wlk. st. a, b, c = Fig. 20, 21, 22, p. 47.
wlk. st. d = crosswise st. (see above).
Yd. = yard: — arms extended horizontally sideways (Fig. 82, p. 90).
Yd. a = Fig. 135 a, p. 124.
Yd. b = see note on p. 123.
Yd. c = Fig. 140, p. 128.
Yd. d = Fig. 141, p. 128.
Yd. e = Fig. 137, p. 125.
½ = half (see above).
2 = double: — indicates that both limbs execute the movement (or position) which follows this figure.